T0173642

KNOW THYSELF

Stephen M. Fleming is a cognitive neuroscientist at University College London. He is a Wellcome/Royal Society Sir Henry Dale Fellow at the Department of Experimental Psychology and Principal Investigator at the Wellcome Centre for Human Neuroimaging, where he leads the Metacognition Group. He lives in London.

KNOW THYSELF

The Science of Self-Awareness

✦

STEPHEN M. FLEMING

JOHN MURRAY

First published in the United States of America in 2021 by Basic Books
First published in Great Britain in 2021 by John Murray (Publishers)
An Hachette UK company

This paperback edition published in 2022

6

Copyright © Stephen M. Fleming 2021

The right of Stephen M. Fleming to be identified as the Author of the Work
has been asserted by him in accordance with the Copyright,
Designs and Patents Act 1988.

Image of Adelson's checkerboard in Chapter 1 used under CC-BY-SA-4.0 license;
image of neural network in Chapter 10 used under CC-BY-SA-3.0 license; all
other art copyright Stephen M. Fleming.

All figures created by the author unless otherwise noted.

Print book interior design by Jeff Williams.

All rights reserved. No part of this publication may be reproduced, stored in a
retrieval system, or transmitted, in any form or by any means without the prior
written permission of the publisher, nor be otherwise circulated in any form of
binding or cover other than that in which it is published and without a similar
condition being imposed on the subsequent purchaser.

A CIP catalogue record for this title is available from the British Library

Paperback ISBN 978-1-529-34506-3
eBook ISBN 978-1-529-34504-9

Typeset in Sabon LT Std

Printed and bound in Great Britain by Clays Ltd, Elcograf S.p.A.

John Murray policy is to use papers that are natural, renewable and
recyclable products and made from wood grown in sustainable forests.
The logging and manufacturing processes are expected to conform
to the environmental regulations of the country of origin.

John Murray (Publishers)
Carmelite House
50 Victoria Embankment
London EC4Y 0DZ

www.johnmurraypress.co.uk

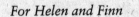

For Helen and Finn

CONTENTS

PREFACE

Imagine you arrive at your doctor's office for an appointment to discuss some recent chest pains. You undergo a series of blood tests and scans, and a week later you return to the clinic, where your doctor reviews the results with you. The condition seems serious, and she briskly recommends surgery for a heart bypass. When you ask your doctor why she is confident the procedure is necessary, she walks you through her thought process, including the possibility that she is wrong and what being wrong might entail, before reiterating her advice that you undergo the surgery. What would you do?

Now imagine that, after you undergo a series of blood tests and scans, the data is fed into an artificially intelligent assistant, which confidently states that the condition seems serious and it would be desirable if you had surgery for a heart bypass. When you ask your doctor whether this is really necessary, she can't tell you; she doesn't know why the recommendation has been made. All she can say is that, when fed the full range of test data, the AI has been highly accurate in the past, and that it would be wise to trust it and proceed with the surgery. What would you do?

In the first case, the answer probably seems obvious: if the doctor is confident and able to explain her reasons for being confident, you feel you should trust her advice. In the second, however, it may not be so clear. Many of us intuitively feel that if a person or a machine is going to be making high-stakes decisions on our behalf, we should be able to ask them to explain *why* they have come up with a particular answer. Many of our legal frameworks—those that ascribe liability and blame for errors—are based on the notion of being able to justify and defend what we did and why we did it. Without an explanation, we are left with blind trust—in each other, or in our machines. Ironically, some of the highest performing machine learning algorithms are often the least explainable. In contrast, humans are rapacious explainers of what we are doing and why, a capacity that depends on our ability to reflect on, think about, and know things about ourselves, including how we remember, perceive, decide, think, and feel.

Psychologists have a special name for this kind of self-awareness: metacognition—literally, the ability to think about our own thinking, from the Greek "meta" meaning "after" or "beyond." Metacognition is a fragile, beautiful, and frankly bizarre feature of the human mind, one that has fascinated scientists and philosophers for centuries. In the biologist Carl Linnaeus's famous 1735 book *Systema Naturae*, he carefully noted down the physical features of hundreds of species. But when it came to our genus, *Homo*, he was so captivated with humans' ability for metacognition that he simply annotated his entry with the one-line Latin description "Nosce te ipsum"—those that know themselves.[1]

Self-awareness is a defining feature of human experience. Take a student, Jane, who is studying for an engineering exam. What might be going through her head? She is no doubt juggling a range of facts and formulas that she needs to master and understand. But she is also, perhaps without realizing it, figuring out how, when, and what to study. Which environment is better, a lively coffee shop or a quiet library? Does she learn best by rereading her notes or by practicing problem sets? Would it be better to shut the book on one topic and move onto another? Can she stop studying altogether and head out with friends?

Getting these decisions right is clearly critical for Jane's chances of success. She would not want to fall into a trap of thinking she knows a topic well when she does not, or to place her trust in a dodgy study strategy. But no one is giving her the answers to these questions. Instead, she is relying on her awareness of how she learns.

Our powers of self-reflection do not lose their importance when we leave the classroom or the exam hall. Consider the experience of James Nestor, an author and free diver. In his book *Deep*, Nestor recounts how he traveled to coastal locations in Greece and the Bahamas to report on free-diving tournaments. At each tournament, there is only one goal: to dive deeper than all the other competitors, all on a single breath. To prove that they have reached a particular depth, the divers retrieve a tag with a number emblazoned on it. If they pass out after surfacing, the dive is declared null and void. To be successful, professional free divers must be acutely self-aware of their ability to reach a depth while avoiding injury or even death. Slight underconfidence will lead to underperformance, whereas slight overconfidence may be fatal. It's telling that a large part of free divers' training takes place on land, in psychological exploration of their underwater capacities and limitations.[2]

Or how about the case of Judith Keppel, one of the first contestants on the British TV game show *Who Wants to Be a Millionaire?* For each question, contestants are asked if they are sure they know the right answer and want to risk their existing winnings on the chance of a higher prize, or if they'd prefer to walk away with whatever they have already won. The stakes are high: being wrong means losing everything you have earned. In

Keppel's case, she faced this decision with £500,000 on the line. The million-pound question was: "Which king was married to Eleanor of Aquitaine?" After a brief discussion with the show's host, Chris Tarrant, she settled on the answer of Henry II. Then Tarrant asked his killer question, the moment when contestants typically agonize the most: "Is that your final answer?" Success again rests on self-awareness. You want to know if you're likely to be right before accepting the gamble. Keppel stuck to her guns and became the show's first winner of the top prize.

What unites the stories of Jane, James, and Judith is how keenly their success or failure hinges on having accurate self-awareness. To appreciate the power of metacognition, we can reimagine their stories in a world where self-awareness is inaccurate. Jane might have erroneously thought that because the fluid mechanics problems felt easy, she could close the book on that topic and move on. She would think she was doing fine, even if this was not the case. A metacognitive error such as this could lead to catastrophic failure in the exam, despite Jane's raw ability and diligent studying. In Judith's case, we can identify two types of metacognitive failure: She may have known the answer but thought she did not, and therefore would have missed out on the opportunity to become a millionaire. Or she may have been overconfident, choosing to gamble on a wrong answer and losing everything. In James's case, such overconfidence may even be the difference between life and death. If he had thought that he was able to handle deeper depths than he was capable of, he would, like a submarine Icarus, have overreached and realized his mistake only when it was too late.

We often overlook the power of metacognition in shaping our own lives, both for good and ill. The relevance of good self-awareness can seem less obvious than, say, the ability to solve equations, perform athletic feats, or remember historical facts. For most of us, our metacognition is like the conductor of an orchestra, occasionally intervening to nudge and guide the players in the right (or wrong) direction in ways that are often unnoticed or unappreciated at the time. If the conductor was absent, the orchestra would continue to play—just as Jane, James, and Judith would continue to plow on with studying, diving, and game-show answering even if their self-awareness was temporarily abolished. But a good conductor can make the difference between a routine rehearsal and a world-class performance—just as the subtle influence of metacognition can make the difference between success and failure, or life and death.

Another reason why the role of self-awareness is sometimes ignored is that it has historically proven difficult to measure, define, and study. But this is now changing. A new branch of neuroscience—metacognitive neuroscience—is pulling back the veil on how the human mind self-reflects. By combining innovative laboratory tests with the latest in brain imaging technology, we are now gaining an increasingly detailed picture of how self-awareness works, both as a cognitive process and as a biological one. As we will see, a science of metacognition can take us further than ever before in knowing ourselves.[3]

Creating a Science of Self-Awareness

I have been fascinated by the puzzle of self-awareness ever since I was a teenager, when I was drawn to books on the brain and mind. I remember glancing up from one of those books while lying by a pool during a summer vacation and daydreaming about my experience: Why should the mere activity of brain cells in my head lead to *this* unique experience of light shimmering on the surface of the swimming pool? And more to the point: How can the very same brain that is having this experience allow me to think about these mysteries in the first place? It was one thing to be conscious, but to know I was conscious and to think about my own awareness—that's when my head began to spin. I was hooked.

I now run a neuroscience lab dedicated to the study of self-awareness at University College London. My team is one of several working within the Wellcome Centre for Human Neuro-imaging, located in an elegant town house in Queen Square in London.[4] The basement of our building houses large machines for brain imaging, and each group in the Centre uses this technology to study how different aspects of the mind and brain work: how we see, hear, remember, speak, make decisions, and so on. The students and postdocs in my lab focus on the brain's capacity for self-awareness. I find it a remarkable fact that something unique about our biology has allowed the human brain to turn its thoughts on itself.

Until quite recently, however, this all seemed like nonsense. As the nineteenth-century French philosopher Auguste Comte put it: "The thinking individual cannot cut himself in two—one of the parts reasoning, while the other is looking on. Since in this case the organ observed and the observing organ are identical, how could any observation be made?"[5] In other words, how can the same brain turn its thoughts upon itself?

Comte's argument chimed with scientific thinking at the time. After the Enlightenment dawned on Europe, an increasingly popular view was that self-awareness was special and not something that could be studied using the tools of science. Western philosophers were instead using self-reflection as a philosophical tool, much as mathematicians use algebra in the pursuit of new mathematical truths. René Descartes relied on self-reflection in this way to reach his famous conclusion "I think, therefore I am," noting along the way that "I know clearly that there is nothing that can be perceived by me more easily or more clearly than my own mind." Descartes proposed that a central soul was the seat of thought and reason, commanding our bodies to act on our behalf. The soul could not be split in two—it just *was*. Self-awareness was therefore mysterious and indefinable, and off-limits to science.[6]

We now know that the premise of Comte's worry is false. The human brain is not a single, indivisible organ. Instead, the brain is made up of billions of small components—neurons—that each crackle with electrical activity and participate in a wiring diagram of mind-boggling complexity. Out of the inter-actions among these cells, our entire mental life—our thoughts and feelings, hopes and dreams—flickers in and out of existence.

But rather than being a meaningless tangle of connections with no discernible structure, this wiring diagram also has a broader architecture that divides the brain into distinct regions, each engaged in specialized computations. Just as a map of a city need not include individual houses to be useful, we can obtain a rough overview of how different areas of the human brain are working together at the scale of regions rather than individual brain cells. Some areas of the cortex are closer to the inputs (such as the eyes) and others are further up the process-ing chain. For instance, some regions are primarily involved in seeing (the visual cortex, at the back of the brain), others in pro-cessing sounds (the auditory cortex), while others are involved in storing and retrieving memories (such as the hippocampus).

In a reply to Comte in 1865, the British philosopher John Stuart Mill anticipated the idea that self-awareness might also depend on the interaction of processes operating within a single brain and was thus a legitimate target of scientific study. Now, thanks to the advent of powerful brain imaging technologies such as functional magnetic resonance imaging (fMRI), we know that when we self-reflect, particular brain networks indeed crackle into life and that damage or disease to these same networks can lead to devastating impairments of self-awareness.[7]

Know Thyself Better

I often think that if we were not so thoroughly familiar with our own capacity for self-awareness, we would be gobsmacked that the brain is able to pull off this marvelous conjuring trick. Imagine for a moment that you are a scientist on a mission to study new life-forms found on a distant planet. Biologists back on Earth are clamoring to know what they're made of and what makes them tick. But no one suggests just asking them! And yet a Martian landing on Earth, after learning a bit of English or Spanish or French, could do just that. The Martians might be stunned to find that we can already tell them something about what it is like to remember, dream, laugh, cry, or feel elated or regretful—all by virtue of being self-aware.[8]

But self-awareness did not just evolve to allow us to tell each other (and potential Martian visitors) about our thoughts and feelings. Instead, being self-aware is central to how we experience the world. We not only perceive our surroundings; we can also reflect on the beauty of a sunset, wonder whether our vision is blurred, and ask whether our senses are being fooled by illusions or magic tricks. We not only make decisions about whether to take a new job or whom to marry; we can also reflect on whether we made a good or bad choice. We not only recall childhood memories; we can also question whether these memories might be mistaken.

Self-awareness also enables us to understand that other people have minds like ours. Being self-aware allows me to ask, "How does this seem to me?" and, equally importantly, "How will this seem to someone else?" Literary novels would become meaningless if we lost the ability to think about the minds of others and compare their experiences to our own. Without self-awareness, there would be no organized education. We would not know who needs to learn or whether we have the capacity to teach them. The writer Vladimir Nabokov elegantly captured this idea that self-awareness is a catalyst for human flourishing:

> Being aware of being aware of being. In other words, if I not only know that I *am* but also know that I know it, then I belong to the human species. All the rest follows—the glory of thought, poetry, a vision of the universe. In that respect, the gap between ape and man is immeasurably greater than the one between amoeba and ape.[9]

In light of these myriad benefits, it's not surprising that cultivating accurate self-awareness has long been considered a wise and noble goal. In Plato's dialogue *Charmides*, Socrates has just returned from fighting in the Peloponnesian War. On his way home, he asks a local boy, Charmides, if he has worked out the meaning of *sophrosyne*—the Greek word for temperance or moderation, and the essence of a life well lived. After a long debate, the boy's cousin Critias suggests that the key to *sophrosyne* is simple: self-awareness. Socrates sums up his argument: "Then the wise or temperate man, and he only, will know himself, and be able to examine what he knows or does not know. . . . No other person will be able to do this."[10]

Likewise, the ancient Greeks were urged to "know thyself" by a prominent inscription carved into the stone of the Temple of Delphi. For them, self-awareness was a work in progress and something to be striven toward. This view persisted into medieval religious traditions: for instance, the Italian priest and philosopher Saint Thomas Aquinas suggested that while God knows Himself by default, we need to put in time and effort to know our own minds. Aquinas and his monks spent long hours engaged in silent contemplation. They believed that only by participating in concerted self-reflection could they ascend toward the image of God.[11]

A similar notion of striving toward self-awareness is seen in Eastern traditions such as Buddhism. The spiritual goal of enlightenment is to dissolve the ego, allowing more transparent and direct knowledge of our minds acting in the here and now. The founder of Chinese Taoism, Lao Tzu, captured this idea that gaining self-awareness is one of the highest pursuits when he wrote, "To know that one does not know is best; Not to know but to believe that one knows is a disease."[12]

Today, there is a plethora of websites, blogs, and self-help books that encourage us to "find ourselves" and become more self-aware. The sentiment is well meant. But while we are often urged to have better self-awareness, little attention is paid to how self-awareness actually works. I find this odd. It would be strange to encourage people to fix their cars without knowing how the engine worked, or to go to the gym without knowing which muscles to exercise. This book aims to fill this gap. I don't pretend to give pithy advice or quotes to put on a poster. Instead, I aim to provide a guide to the building blocks of self-awareness, drawing on the latest research from psychology, computer science, and neuroscience. By understanding how self-awareness works, I aim to put us in a position to answer the Athenian call to use it better.

I also aim to help us use our machines better—both those that exist today and those that are likely to arrive in the near future. As with your imagined visit to the doctor's artificially intelligent clinic and its inexplicable advice to have surgery, we are already being forced to deal with complex systems making decisions we do not understand. We are surrounded by smart but unconscious algorithms—from climate forecasting models to automatic financial traders—and similar tools are poised to encroach on all areas of our lives. In many cases, these algorithms make our lives easier and more productive, and they may be required to help us tackle unprecedented challenges such as climate change. But there is also a danger that defer-ring to supersmart black boxes will limit human autonomy: by removing metacognition from the equation, we will not know why or how certain decisions were made and instead be forced into blindly following the algorithms' advice. As the philoso-pher Daniel Dennett points out: "The real danger, I think, is not that machines more intelligent than we are will usurp our role as captains of our destinies, but that we will overestimate the comprehension of our latest thinking tools, prematurely ceding authority to them far beyond their competence."[13] As we will see, the science of self-awareness provides us with alternative visions of this future, ones that ensure that awareness of com-petence remains at the top of the priority list, both for ourselves and our machines.

Looking Ahead

Let's take a look at the road ahead. The big idea of this book is that the human brain plays host to specific algorithms for self-awareness. How these algorithms work will occupy us in Part I. We will see that the neural circuits supporting metacog-nition did not just pop up out of nowhere. Instead, they are grounded in the functions of the evolved human brain. This means that many of the building blocks of metacognition are also shared with other species and are in place early in human development. We'll cover both the unconscious processes that form the building blocks of self-monitoring and the conscious processes that enable you to be self-aware of the experiences you are having. As will become clear, when we talk about self-awareness, what we really mean is a collection of capaci-ties—such as being able to recognize our mistakes and comment on our experience—that when bundled together result in a self-aware human being.[14]

By the end of Part I, we will have seen how a number of critical components come together to create a fully-fledged capacity for self-awareness. We will also be in a position to understand how and why these processes sometimes go wrong, leading to failures of self-awareness in diseases such as schizophrenia and dementia. In Part II, we will then turn to how we use self-awareness in many areas of our lives to learn, make decisions, and collaborate with others. By understanding how and why self-awareness may become distorted—and by recognizing both its power and fragility—we will be in a position to ensure that we do not end up in situations in which it tends to misfire. We'll dig into several important arenas of human affairs—including the crucial role that metacognition plays in witnesses testimony, in politics, and in science—to see why knowing ourselves, and knowing how others know themselves, is crucial to building a fairer and better society. We'll explore how self-awareness helps us separate reality from imagination and how, by learning to harness it, it can even help us shape our dreams. We will see that because self-awareness is sometimes absent there are, in fact, plenty of cases in which we humans are also no better than black boxes, unable to explain what we have done or why.

We will also see that, despite these limitations, the human capacity for self-awareness and self-explanation is what underpins our notions of autonomy and responsibility. We'll explore the role of self-awareness in classroom learning and teaching. We'll see why in sports it might be better to be less self-aware to perform well but more self-aware when coaching others. We'll see how digital technology changes our awareness of ourselves and others in a range of crucial ways. Indeed, I'll make the case that in a world of increasing political polarization and misinformation, cultivating the ability to self-reflect and question our beliefs and opinions has never been more essential. We'll explore why computers—even the most powerful—currently lack metacognition, and how the increasing prevalence of machine learning in AI means that algorithms for intelligence are rapidly diverging from algorithms for self-awareness. We'll examine what this might mean for society and how we might fix it, either by attempting to build self-awareness into our computers or by ensuring we can understand and use the machines that we build. However that effort concludes, it may hold the key to solving some of the most pressing problems in society.

By the end of all this, I hope it will be clear why, from ancient Athens to the boardroom of Amazon.com, cultivating self-awareness has always been essential to flourishing and success. But we are getting ahead of ourselves. To unravel the mysteries of how self-awareness works, we need to start with the simplest of building blocks. Let's begin with two features of how our minds work: how we track uncertainty and how we monitor our actions. These two features may appear simple, but they are fundamental components of a self-aware brain.

PART I

✦

BUILDING MINDS
THAT KNOW THEMSELVES

1

HOW TO BE UNCERTAIN

> The other fountain [of] ideas, is the perception of the operation of our own minds within us. . . . And though it be not sense, as having nothing to do with external objects, yet it is very like it, and might properly enough be called internal sense.
>
> **—JOHN LOCKE,**
> *Essay Concerning Human Understanding, Book II*

Is something there, or not? This was the decision facing Stanislav Petrov one early morning in September 1983. Petrov was a lieutenant colonel in the Soviet Air Defense Forces and in charge of monitoring early warning satellites. It was the height of the Cold War between the United States and Russia, and there was a very real threat that long-range nuclear missiles could be launched by either side. That fateful morning, the alarms went off in Petrov's command center, alerting him that five US missiles were on their way to Russia. Under the doctrine of mutually assured destruction, his job was to immediately report the attack to his superiors so they could launch a counterattack. Time was of the essence—within twenty-five minutes, the missiles would detonate on Soviet soil.[1]

But Petrov decided that the alert was unlikely to be a real missile. Instead, he called in a system malfunction. To him, it seemed more probable that the satellite was unreliable—that the blip on the radar screen was noise, not signal—than that the United States had sent over missiles in a surprise attack that would surely launch a nuclear war. After a nervous wait of several minutes, he was proved right. The false alarm had been triggered by the satellites mistaking the sun's reflection off the tops of clouds for missiles scudding through the upper atmosphere.

Petrov saw the world in shades of gray and was willing to entertain uncertainty about what the systems and his senses were telling him. His willingness to embrace ambiguity and question what he was being told arguably saved the world from disaster. In this chapter, we will see that representing uncertainty is a key ingredient in our recipe for creating self-aware systems. The human brain is in fact an exquisite uncertainty-tracking machine, and the role of uncertainty in how brains work goes much deeper than the kind of high-stakes decision facing Petrov. Without an ability to estimate uncertainty, it is unlikely that we would be able to perceive the world at all—and a wonderful side benefit is that we can also harness it to doubt ourselves.

Inverse Problems and How to Solve Them

The reason Petrov's decision was difficult was that he had to separate out signal from noise. The same blip on the radar screen could be due to an actual missile or noise in the system. It is impossible to work out which from the characteristics of the blip alone. This is known as an inverse problem—so called because solving it requires inverting the causal chain and making a best guess about what is causing the data we are receiving. In the same way, our brains are constantly solving inverse problems, unsure about what is really out there in the world.

The reason for this is that the brain is locked inside a dark skull and has only limited contact with the outside world through the lo-fi information provided by the senses. Take the seemingly simple task of deciding whether a light was just flashed in a darkened room. If the light flash is made dim enough, then sometimes you will say the light is present even when it is absent. Because your eye and brain form a noisy system, the firing of neurons in your visual cortex is not exactly the same for each repetition of the stimulus. Sometimes, even when the light isn't flashed, random noise in the system will lead to high firing rates, just like the blip on Petrov's radar screen was caused by atmospheric noise. Because the brain doesn't know whether these high firing rates are caused by signal or noise, if your visual cortical neurons are firing vigorously it will seem as though a light was flashed even if it wasn't.[2]

Because our senses—touch, smell, taste, sight, and hearing—each have access to only a small, noisy slice of reality, they must pool their resources to come up with a best guess about what is really out there. They are rather like the blind men in the ancient Indian parable. The one holding the elephant's leg says it must be a pillar; the one who feels the tail says it is like a rope; the one who feels the trunk says it is like a tree branch; the one who feels the ear says it is like a hand fan; the one who feels its belly says it is like a wall; and the one who feels the tusk says it is like a solid pipe. Eventually, a stranger wanders past and informs them that they are, in fact, all correct and the elephant has all the features they observed. They would do better to combine their perspectives, he says, rather than argue.

A mathematical framework known as Bayes's theorem provides a powerful tool for thinking about these kinds of problems. To see how Bayes's rule helps us solve inverse problems, we can play the following game. I have three dice, two of which are regular dice with the numbers 1 to 6 on their faces, and one of which is a trick die with either a 0 or 3 on every face.

From behind a curtain, I'm going to roll all three dice at once and tell you the combined total. On each roll, I might choose to use a trick die that shows all 0s, or a trick die that shows all 3s. For instance, on my first roll I might roll 2, 4, and 0 (on the third, trick die) for a combined total of 6. Your task is to tell me your best guess about the identity of the trick die—either a 3 or a 0—based only on your knowledge of the total.[3]

In this game, the 0 or the 3 on the trick die stand in for the "hidden" states of the world: whether the missile was present in Petrov's dilemma, or whether the light was flashed in the case of our visual cortex neuron. Somehow, we need to go back from the noisy evidence we have received—the sum total of the three dice—and use this to work out the hidden state.

Sometimes this is easy. If I tell you the combined total is 4 or less, then you know that the third die must have been showing 0 to produce such a low sum. If the combined total is greater than 12 (two 6s plus a quantity more than 0), then you know for sure that the third die must have been showing 3. But what about quantities between these extremes? What about a total of 6 or 8? This is trickier.

One way we might go about solving this game is by trial and error. We could roll the three dice ourselves many times, record the total, and observe the true state of the world: what was actually showing on the face of the third die on each roll. The first few rolls of the game might look like this:

Roll	Die 1	Die 2	Trick Die	Total
1	2	4	0	6
2	5	1	3	9
3	5	6	3	14

And so on, for many tens of rolls. An easier way to present this data is in a chart of the number of times we observe a particular total—say, 6—and the identity of the trick die at the time (0 or 3). We can select particular colors for the trick die number; here I've chosen gray for 0 and white for 3.

After ten rolls the graph might look like this:

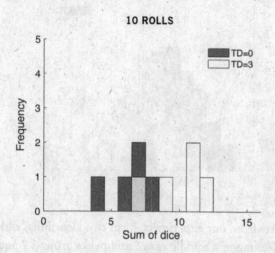

This isn't very informative, and shows only a scatter of different totals, just like in our table. But after fifty rolls a pattern starts to emerge:

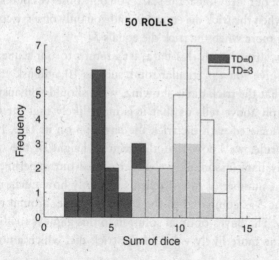

And after one thousand rolls, the pattern is very clear:

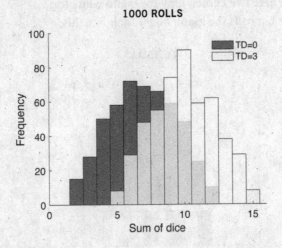

The counts from our experiment form two clear hills, with the majority falling in a middle range and peaks around 7 and 10. This makes sense. On average, the two real dice will give a total of around 7, and therefore adding either 0 or 3 from the trick die to this total will tend to give 7 or 10. And we see clear evidence for our intuition at the start: you only observe counts of 4 or less when the trick die equals 0, and you only observe counts of 13 or more when the trick die equals 3.

Now, armed with this data, let's return to our game. If I were to give you a particular total, such as 10, and ask you to guess what the trick die is showing, what should you answer? The graph above tells us that it is more likely that the total 10 is associated with the trick die having 3 on its face. From Bayes's rule, we know that the relative height of the white and gray bars (assuming we've performed our experiment a sufficient number of times) tells us precisely how much more likely the 3 is compared to the 0—in this case, around twice as likely. The Bayes-optimal solution to this game is to always report the more likely value of the trick die, which amounts to saying 3 when the total is 9 or above and 0 when the total is 8 or lower.

What we've just sketched is an algorithm for making a decision from noisy information. The trick die is always lurking in the background because it is contributing to the total each time. But its true status is obscured by the noise added by the two normal dice, just as the presence of a missile could not be estimated by Petrov from the noisy radar signal alone. Our game is an example of a general class of problems involving decisions under uncertainty that can be solved by applying Bayes's rule.

In the case of Petrov's fateful decision, the set of potential explanations is limited: either there is a missile or it's a false alarm. Similarly, in our dice game, there are only two explanations to choose between: either the trick die is a 0 or it's a 3. But in most situations, not only is the sensory input noisy, but there is a range of potential explanations for the data streaming in through our senses. Imagine a drawing of a circle around twenty centimeters across and held at a distance of one meter from the eye. Light reflected from the circle travels in straight lines, passing through the lens of the eye and creating a small image (of a circle) on the retina. Because the image on the retina is two-dimensional, the brain could interpret it as being caused by any infinite number of circles of different sizes arranged at appropriate distances. Roughly the same retinal image would be caused by a forty-centimeter circle held at two meters, or an eight-meter circle at forty meters. In many cases, there is simply not enough information in the input to constrain what we see.

These more complex inverse problems can be solved by making guesses as to the best explanation based on additional information drawn from other sources. To estimate the actual diameter of the circle, for instance, we can use other cues such as differences in the images received by the two eyes, changes in the texture, position, and shading of nearby objects, and so on.

To experience this process in real time, take a look at these two pictures:

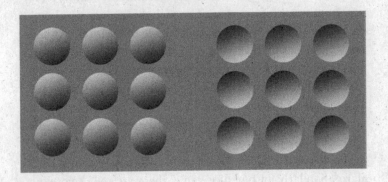

Most people see the image on the left-hand side as a series of bumps, raised above the surface. The image on the right, in contrast, looks like a series of little pits or depressions in the page. Why the difference?

The illusion is generated by your brain's solution to the inverse problem. The left and right sets of dots are actually the same image rotated 180 degrees (you can rotate the book to check!). The reason they appear different is that our visual system expects light to fall from above, because scenes are typically lit from sunlight falling from above our heads. In contrast, uplighting—such as when light from a fire illuminates the side of a cliff, or spotlights are projected upward onto a cathedral—is statistically less common. When viewing the two sets of dots, our brain interprets the lighter parts of the image on the left as being consistent with light striking a series of bumps and the darker parts of the image on the right as consistent with a series of shadows cast by holes, despite both being created from the same raw materials.

Another striking illusion is this image created by the vision scientist Edward Adelson:

Adelson's checkerboard *(Edward H. Adelson)*

In the left-hand image, the squares labeled A and B are in fact identical shades of gray; they have the same luminance. Square B appears lighter because your brain "knows" that it is in shadow: in order to reflect the same level of light to the eye as A, which is fully illuminated, it must have started out lighter. The equivalence of A and B can be easily appreciated by connecting them up, as in the right-hand image. Now the cue provided by this artificial bridge overrides the influence of the shadow in the brain's interpretation of the squares (to convince yourself that the left- and right-hand images are the same, try using a sheet of paper to cover up the bottom half of the two images).

The upshot is that these surprising illusions are not really illusions at all. One interpretation of the image is given by our scientific instruments—the numbers produced by light meters and computer monitors. The other is provided by our visual systems that have been tuned to discover regularities such as shadows or light falling from above—regularities that help them build useful models of the world. In the real world, with light and shade and shadows, these models would usually be right.

Many visual illusions are clever ways of exposing the workings
of a system finely tuned for perceptual inference. And, as we
will see in the next section, several principles of brain organization are consistent with this system solving inverse problems on
a massive scale.

Building Models of the World

One of the best-understood parts of the human and monkey
brain is the visual system. Distinct regions toward the back of the
brain process different aspects of visual input, and each is labeled
with increasing numbers for more advanced stages of image processing. V1 and V2 extract information about the orientation of
lines and shapes, V4 about color, and V5 about whether objects
are moving. Downstream of these V regions we hit regions of the
ventral visual stream that are tasked with putting all these pieces
together to identify whole objects, such as faces and bodies and
tables and chairs. In parallel, the dorsal visual stream contains
regions that specialize in keeping track of where things are and
whether they are moving from place to place.[4]

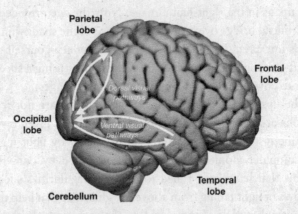

The right hemisphere of the human brain. The locations of the four
cortical lobes, the cerebellum, and key visual pathways are indicated.

At the start of the ventral visual stream, individual brain cells encode only a small amount of the external world, such as a patch in the lower left of our field of view. But as we move up the hierarchy, the cells begin to widen their focus, similar to a camera zooming out. By the time we reach the top of the hierarchy, where a stimulus is displayed matters much less than what it depicts— a face, house, cat, dog, etc. The lens is completely zoomed out, and information about the object's identity is represented independently of spatial location.

Crucially, however, information in the visual system does not just flow in one direction. For a long time, the dominant view of information processing in the brain was that it was a feed-forward system—taking in information from the outside world, processing it in hidden, complex ways, and then spitting out commands to make us walk and talk. This model has now been superseded by a raft of evidence that is difficult to square with the input-output view. In the visual system, for instance, there are just as many, if not more, connections in the reverse direction, known as feedback or top-down connections. Information travels both forward and backward; upper levels of the hierarchy both receive inputs from lower levels and send information back down in constant loops of neural activity. This way of thinking about the mind is known as predictive processing, and it represents a radically different understanding of what the brain does—although one with a long intellectual history, as the range of references in the endnote makes clear.[5]

Predictive processing architectures are particularly well suited to solving inverse problems. Instead of just passively taking in information, the brain can harness these top-down connections to actively construct our perception of the external world and shape what we see, hear, think, and feel. Higher levels furnish information about the type of things we might encounter in any given situation and the range of hypotheses we might entertain. For instance, you might know that your friend owns a Labrador, and so you expect to see a dog when you walk into the house but don't know exactly where in your visual field the dog will appear. This higher-level prior—the spatially invariant concept of "dog"—provides the relevant context for lower levels of the visual system to easily interpret dog-shaped blurs that rush toward you as you open the door.

The extent to which our perceptual systems should rely on these regularities—known as priors—is in turn dependent on how uncertain we are about the information being provided by our senses. Think back to Petrov's dilemma. If he was sure that his missile-detection technology was flawless and never subject to error, he would have been less willing to question what the system was telling him. Whether we should adjust our beliefs upon receiving new data depends on how reliable we think that information is.

In fact, Bayesian versions of predictive processing tell us that we should combine different sources of information—our prior beliefs and the data coming in through our senses—in inverse proportion to how uncertain we are about them. We can think of this process as being similar to pouring cake batter into a flexible mold. The shape of the mold represents our prior assumptions about the world. The batter represents the sensory data—the light and sound waves hitting the eyes and ears. If the incoming data is very precise or informative, then the batter is very thick, or almost solid, and will be hardly affected by the shape of the mold (the priors). If, in contrast, the data is less precise, then the batter will be runnier, and the shape of the mold will dominate the shape of the final product.

For instance, our eyes provide more precise information about the location of objects than our hearing. This means that vision can act as a useful constraint on the location of a sound source, biasing our perception of where the sound is coming from. This is used to great effect by ventriloquists, who are seemingly able to throw their voices to a puppet held at arm's length. The real skill of ventriloquism is the ability to speak without moving the mouth. Once this is achieved, the brains of the audience do the rest, pulling the sound to its next most likely source, the talking puppet.[6]

It makes sense, then, that keeping track of uncertainty is an inherent part of how the brain processes sensory information. Recordings of cells from the visual cortex show us how this might be done. It's well known that moving objects such as a waving hand or a bouncing ball will activate neurons in an area of the monkey brain known as MT (the human equivalent is V5). But cells in MT do not just activate for any direction of movement. Some cells fire most strongly for objects moving to the left, others for up, down, and all other points of the compass. When firing rates of MT cells are recorded over multiple presentations of different motion directions, they begin to form a distribution like the ones we saw in our dice game. At any moment in time, these populations of MT cells can be thought of as signaling the uncertainty about a particular direction of motion, just as our noisy dice total signaled the probability of the trick die being a 0 or a 3.[7]

Uncertainty is also critical for estimating the states of our own bodies. Information about where our limbs are in space, how fast our heart is beating, or the intensity of a painful stimulus is conveyed to the skull by sensory neurons. From the brain's perspective, there is little difference between the electrical impulses traveling down the optic nerve and the neural signals ascending from our gut, heart, muscles, or joints. They are all signals of what might be happening outside of the skull, and these signals are subject to illusions of the kind that we encountered for vision. In one famous experiment, stroking a rubber hand in time with the participant's own (hidden) hand is sufficient to convince the participant that the rubber hand is now their own. In turn, the illusion of ownership of the new rubber hand leads the brain to wind down the neural signals being sent to the actual hand. Just as the voice is captured by the ventriloquist's dummy, the synchrony with which the rubber hand is seen and felt to be stroked pulls the sense of ownership away from the real hand.[8]

Harnessing Uncertainty to Doubt Ourselves

Of course, no one is suggesting that we consciously churn through Bayesian equations every time we perceive the world. Instead, the machinery our brains use to solve inverse problems is applied without conscious thought, in what the German physicist Hermann von Helmholtz called a process of "unconscious inference." Our brains rapidly estimate the effects of light and shade on the dips, bumps, and checkerboards we encountered in the images on previous pages, literally in the blink of an eye. In a similar fashion, we reconstruct the face of a close friend, the taste of a fine wine, and the smell of freshly baked bread by combining priors and data, carefully weighting them by their respective uncertainties. Our perception of the world is what the neuroscientist Anil Seth refers to as a "controlled hallucination"—a best guess of what is really out there.

It is clear that estimating uncertainty about various sources of information is fundamental to how we perceive the world. But there is a remarkable side benefit of these ingenious solutions to the inverse problem. In estimating uncertainty in order to perceive the world, we also gain the ability to doubt what we perceive. To see how easy it is to turn uncertainty into self-doubt, let's consider the dice game again. As the numbers in the game tend toward either 15 or 0, we become surer about the trick die showing a 3 or 0, respectively. But in the middle part of the graph, where the gray and white bars are of similar height—totals of 7 or 8—there is limited support for either option. If I ask you how confident you are about your response, it would be sensible to doubt decisions about the numbers 7 and 8 and to be more confident about smaller and larger quantities. In other words, we know that we are likely to know the answer when uncertainty is low, and we know that we are likely to *not* know the answer when uncertainty is high.

Bayes's rule provides us with a mathematical framework for thinking about these estimates of uncertainty, sometimes known as type 2 decisions—so called because they are decisions about the accuracy of other decisions, rather than type 1 decisions, which are about things in the world. Bayes's theorem tells us that it is appropriate to be more uncertain about responses toward the middle of the graph, because they are the ones most likely to result in errors and are associated with the smallest probability of being correct. Conversely, as we go out toward the tails of each distribution, the probability of being correct goes up. By harnessing the uncertainty that is inherent to solving inverse problems, we gain a rudimentary form of metacognition for free—no additional machinery is needed.[9]

And, because tracking uncertainty is foundational to how brains perceive the world, it is not surprising that this form of metacognition is widespread among a range of animal species. One of the first—and most ingenious—experiments on animal metacognition was developed by the psychologist J. David Smith in his study of a bottlenose dolphin named Natua. Smith trained Natua to press two different levers in his tank to indicate whether a sound was high-pitched or low-pitched. The low-pitched sound varied in frequency from very low to relatively high, almost as high as the high-pitched sound. There was thus a zone of uncertainty in which it wasn't always clear whether low or high was the right answer, just like in our dice game.[10]

Once Natua had got the hang of this task, a third lever was introduced into the tank that could be pressed to skip the current trial and move on to the next one—the dolphin equivalent of skipping a question on a multiple-choice quiz. Smith reasoned that if Natua declined to take on decisions when his uncertainty about the answer was high, he would be able to achieve a higher accuracy overall than if he was forced to guess. And this is exactly what Smith found. The data showed that Natua pressed the third lever mostly when the sound was ambiguous. As Smith recounts, "When uncertain, the dolphin clearly hesitated and wavered between his two possible responses, but when certain, he swam towards his chosen response so fast that his bow wave would soak the researchers' electronic switches."[11]

Macaque monkeys—which are found across Asia (and are fond of stealing tourists' food at temples and shrines)—also easily learn to track their uncertainty in a similar setup. In one experiment, macaques were trained to judge which was the biggest shape on a computer screen, followed by another choice between two icons. One icon led to a risky bet (three food pellets if right, or the removal of food if wrong), while the other, safe option always provided one food pellet—the monkey version of *Who Wants to Be a Millionaire?* The monkeys selected the risky option more often when they were correct, a telltale sign of metacognition. Even more impressively, they were able to immediately transfer these confidence responses to a new memory test without further training, ruling out the idea that they had just learned to associate particular stimuli with different confidence responses. Adam Kepecs's lab, based at Cold Spring Harbor in New York, has used a version of this task to show that rats also have a sense of whether they are likely to be right or wrong about which of two perfumes is most prominent in a mixture of odors. There is even some evidence to suggest that birds can transfer their metacognitive competence between different tests, just like monkeys.[12]

If a sensitivity to uncertainty is a fundamental property of how brains work, it makes sense that this first building block of metacognition might also be found early in the lives of human babies. Taking inspiration from Smith's tests, Louise Goupil and Sid Kouider at the École Normale Supérieure in Paris set out to measure how eighteen-month-old infants track uncertainty about their decisions. While sitting on their mothers' laps, the babies were shown an attractive toy and allowed to play with it to whet their appetite for more playtime in the future. They then saw the toy being hidden in one of two boxes. Finally, after a brief delay, they were allowed to reach inside either of the boxes to retrieve the toy.

In reality, the toy was sneakily removed from the box by the experimenter. This allowed the researchers to measure the infants' confidence about their choice of box. They reasoned that, if the babies knew whether they were making a good or bad choice, they would be more willing to search for the (actually nonexistent) toy when the correct box was chosen compared to when they chose incorrectly. This was indeed the case: when babies made wrong moves, they were less persistent in searching for the toy. They were also more likely to ask their mother for help in retrieving the toy when they were most prone to making an error. This data tells us that even at a young age, infants can estimate how uncertain they are about simple choices, asking for help only when they most need it.[13]

We cannot know for sure how animals and babies are solving these problems, because—unlike human adults—they cannot tell us about what they are thinking and feeling. A critic could argue that they are following a lower-level rule that is shared across all the tasks in the experiment—something like, If I take a long time to decide, then I should press the "uncertain" key—without forming any feeling of uncertainty about the decisions they are making. In response to this critique, ever more ingenious experiments have been designed to rule out a variety of non-metacognitive explanations. For instance, to rule out tracking response time, other studies have given the animals the chance to bet on their choices before they have started the test and before response-time cues are available. In this setup, macaque monkeys are more likely to be correct when they choose to take the test than when they decline, suggesting that they know when they know the answer—a hallmark of metacognition.[14]

There is also evidence that otherwise intelligent species fail to track uncertainty in these situations, suggesting that feelings of uncertainty might really be picking up on the first glimmers of self-awareness, rather than a more generic cognitive ability. Capuchin monkeys, a New World species found in South America, share many characteristics with macaques, using tools such as stones to crack open palm nuts, and living in large social groups. But capuchins appear unable to signal that they are uncertain in Smith's task. In a clever twist, it is possible to show that capuchins have no difficulty using a third response key to classify a new stimulus, but they are unable to use the same response to indicate when they are uncertain. This data suggests that when comparing two similar species of monkey, one may show signs of metacognition while another may not.[15]

Once uncertainty tracking is in place, it opens the door to a range of useful behaviors. For starters, being able to estimate uncertainty means we can use it to decide whether or not we need more information. Let's go back to our dice game. If I were to give you a total near the middle of the graph—a 7 or an 8—then you might reasonably be uncertain about whether to answer 0 or 3. Instead, you might ask me to roll the dice again. If I were to then roll a 5, a 4, and a 7, all with the same three dice, then you would be much more confident that the trick die was a 0. As long as each roll is independent of the previous one, Bayes's theorem tells us we can compute the probability that the answer is a 3 or a 0 by summing up the logarithm of the ratio of our confidence in each hypothesis after each individual roll.[16]

The brilliant British mathematician Alan Turing used this trick to figure out whether or not to change tack while trying to crack the German Enigma code in the Second World War. Each morning, his team would try new settings of their Enigma machine in an attempt to decode intercepted messages. The problem was how long to keep trying a particular pair of ciphers before discarding it and trying another. Turing showed that by accumulating multiple samples of information over time, the code breakers could increase their confidence in a particular setting being correct—and, critically, minimize the amount of time wasted testing the wrong ciphers.[17]

In the same way, we can use our current estimate of confidence to figure out whether a new piece of information will be helpful. If I get a 12 on my first roll, then I can be reasonably confident that the trick die is showing a 3 and don't need to ask for the dice to be rolled again. But if I get a 7 or 8, then it would be prudent to roll again and resolve my current uncertainty about the right answer. The role of confidence in guiding people's decisions to seek new information has been elegantly demonstrated in the lab. Volunteers were given a series of difficult decisions to make about the color of shapes on a computer screen. By arranging these shapes in a particular way, the researchers could create conditions in which people *felt* more uncertain about the task but performed no worse. This design nicely isolates the effect a feeling of uncertainty has on our decisions. When asked whether they wanted to see the information again, participants did so only when they felt more uncertain. Just as in the experiments on babies, the participants were relying on internal feelings of uncertainty or confidence to decide whether to ask for help.[18]

Shades of Gray

Being able to track uncertainty is fundamental to how our brains perceive the world. Due to the complexity of our environment and the fact that our senses provide only low-resolution snapshots of our surroundings, we are forced to make assumptions about what is really out there. A powerful approach to solving these inverse problems combines different sources of data according to their reliability or uncertainty. Many aspects of this solution are in keeping with the mathematics of Bayesian inference, although there is a vigorous debate among neuroscientists as to how and whether the brain implements (approximations to) Bayes's rule.[19]

Regardless of how it is done, we can be reasonably sure that computing uncertainty is a fundamental principle of how brains work. If we were unable to represent uncertainty, we would only ever be able to see the world in one particular way (if at all). By representing uncertainty we also acquire our first building block of metacognition—the ability to doubt what our senses are telling us. By itself, the ability to compute uncertainty is not sufficient for full-blown self-awareness. But it is likely sufficient for the rudimentary forms of metacognition that have been discovered in animals and babies. Nabokov's bright line between humans and other species is becoming blurred, with other animals also demonstrating the first signs of metacognitive competence.

But tracking uncertainty is only the beginning of our story. Up until now we have treated the brain as a static perceiver of the world, fixed in place and unable to move around. As soon as we add in the ability to act, we open up entirely new challenges for metacognitive algorithms. Meeting these challenges will require incorporating our next building block: the ability to monitor our actions.

ALGORITHMS FOR SELF-MONITORING

> Once the ability to sense external goings-on has developed, however, there ensues a curious side effect that will have vital and radical consequences. This is the fact that the living being's ability to sense certain aspects of its environment flips around and endows the being with the ability to sense certain aspects of itself.

> **—DOUGLAS HOFSTADTER,** *I Am a Strange Loop*

In the previous chapter, we saw how the perceptual systems of the brain work, taking in information from the senses and solving inverse problems to build a model of the world. We also saw how the ability to encode and track different sources of uncertainty about what we are seeing and hearing is critical to how we perceive and can also be harnessed to doubt ourselves. In the next leg of our tour of self-monitoring algorithms, we are going to start at the opposite end of the system: the parts that control our actions.

We typically think of an action as something conscious and deliberate—preparing dinner, for instance, or picking up the phone to call a friend. But the range of possible actions that we initiate over the course of a normal day is much broader.

Everything that our bodies do to change things in our environment can be considered an action: breathing, digesting, speaking, adjusting our posture. And, as this set of examples illustrates, many of our actions occur unconsciously and are critical for keeping us alive.

The problem with actions, though, is that they do not always go to plan. This means that we need ways of correcting our mistakes quickly and efficiently. Any self-aware system worth its salt would not want to simply fire off actions like darts, with no way of correcting their flight after they have been thrown. Imagine reaching for your glass of red wine at a dinner party, only to misjudge and tip it off the edge of the table. You watch in horror, feeling helpless to do anything about it. And yet, sometimes, if you're lucky, your hand will seem to shoot out of its own accord, catching the glass barely milliseconds into its descent. We will see that at the heart of being able to correct our actions in this way is an ability to *predict* what should have happened—bringing the glass into our hand—but didn't.

In fact, prediction is at the heart of algorithms that can monitor themselves. Consider the predictive text on your smartphone. The only way it can correct ham-fisted errors in your typing is by having some running estimate of what word you intended to say in the first place (and if those estimates are, in fact, at odds with your intentions, then you experience maddeningly persistent sabotage). This is also intuitively true of self-awareness. We can only recognize our errors and regret our mistakes if we know what we should have done but didn't do. The French have a wonderful phrase for this: "L'esprit d'escalier," or staircase wit. These are the things you realize you should have said upon descending the stairs on your way out of a party.

In this chapter, we will discover a panoply of algorithms that predict and correct our actions. As we will see, brains of all shapes and sizes have exquisite fail-safe mechanisms that allow them to monitor their performance—ranging from the smallest of corrective adjustments to the movement of my arm as I reach out to pick up my coffee, all the way to updating how I feel about my job performance. Being able to monitor our actions is our second building block of self-awareness.

Predicting Our Errors

Some of the simplest and most important actions that we engage in are those aimed at maintaining our internal states. All living things must monitor their temperature, nutrient levels, and so on, and doing so is often a matter of life or death. If these states move too far from their ideal position (known as set points), staying alive becomes very difficult indeed. Consider a humble single-celled bacterium. Living cells depend on managing the acidity of their internal world, because most proteins will cease to function beyond a narrow range of pH. Even simple bacteria have intricate networks of sensors and signaling molecules on their cell surface, which lead to the activation of pumps to restore a neutral pH balance when required.

This is known as homeostasis, and it is ubiquitous in biology. Homeostasis works like the thermostat in your house: when the temperature drops below a certain point, the thermostat switches on the heating, ensuring that the ambience of your living room is kept within a comfortable range. A curious aspect of homeostasis is that it is recursive—it seeks to alter the very same thing that it is monitoring. The thermostat in my living room is trying to regulate the temperature of the same living room, not some room in my neighbor's house. This feature of homeostasis is known as a closed-loop system. If the state it is detecting is in an acceptable range, then all is well. If it's not—if an imbalance in pH or temperature is detected—some action is taken, and the imbalance is corrected. Homeostasis can often be left to its own devices when up and running; it is rare that a corrective action will not have a desired effect, and the control process, while intricate, is computationally simple.

Homeostatic mechanisms, however, operate in the here and now, without caring very much about the future. A simple on-off thermostat cannot "know" that it tends to get colder at night and warmer during the day. It just switches on the heating when the temperature drops below a threshold. In the BBC comedy series *Peep Show*, Jez misunderstands this critical feature of thermostats, telling his housemate Mark, "Let's whack [the boiler] up to 29. . . . I don't actually want it to be 29, but you've got to give it something to aim for. It'll get hotter, quicker." Mark replies disdainfully (and accurately): "No it won't, it's either on or off. You set it, it achieves the correct temperature, it switches off." You cannot trick a boiler.

The new breed of learning thermostats, such as the Nest, improves on traditional on-off devices by learning the typical rise and fall in temperature over the course of the day and the preferences of the owner for particular temperatures. A smart thermostat can then anticipate when it needs to switch on to maintain a more even temperature. The reason this is more successful than a good-old-fashioned thermostat is a consequence of a classic proposal in computer science known as the good regulator theorem, which states that the most effective way of controlling a system is to develop an accurate model of that same system. In other words, the more accurate my model of the kind of things that affect the temperature, the more likely I will be able to anticipate when I need to make changes to the heating to keep it within a comfortable range.[1]

The same is true when we move beyond homeostasis to actions that affect the external world. In fact, we can think of all our behavior as a form of elaborate homeostasis, in the sense that many of the things we do are aimed at keeping our internal states within desirable bounds. If I am hungry, I might decide to go and make a sandwich, which makes me feel full again. If I need money to buy ingredients to make a sandwich, I might decide to apply for a job to make money, and so on. This idea—that everything we do in life fits into some grand scheme that serves to minimize the "error" in our internal states—has both its proponents and critics in the field of computational neuroscience. But at least for many of our simpler actions, it provides an elegant framework for thinking about how behavior is monitored and controlled. Let's take a closer look at how this works in practice.[2]

Who Is in Control?

In the same way that there are dedicated sensory parts of the brain—those that handle incoming information from the eyes and ears, for instance—there are also dedicated motor structures that send neural projections down to the spinal cord in order to control and coordinate our muscles. And just as the visual cortex is organized hierarchically, going from input to high-level representations of what is out there in the world, the motor cortex is organized as a descending hierarchy. Regions such as the premotor cortex are involved in creating general plans and intentions (such as "reach to the left"), while lower-level brain areas, such as the primary motor cortex, are left to implement the details. Regions in the prefrontal cortex (PFC) have been suggested to be at the top of both the perceptual and motor hierarchies. This makes sense if we think of the PFC as being involved in translating high-level perceptual representations (the red ball is over there) into high-level action representations (let's pick up the red ball).[3]

One consequence of the hierarchical organization of action is that when we reach for a cup of coffee, we do not need to consciously activate the sequence of muscles to send our arm and hand out toward the cup. Instead, most action plans are made at a higher level—we want to taste the coffee, and our arm, hand, and mouth coordinate to make it so. This means that in a skilled task such as playing the piano, there is a delicate ballet between conscious plans unfolding further up the hierarchy (choosing how fast to play, or how much emphasis to put on particular passages) and the automatic and unconscious aspects of motor control that send our fingers toward the right keys at just the right time. When watching a concert pianist at work, it seems as though their hands and fingers have a life of their own, while the pianist glides above it all, issuing commands from on high. As the celebrated pianist Vladimir Horowitz declared, "I am a general, my soldiers are the keys." In the more prosaic language of neuroscience, we offload well-learned tasks to unconscious, subordinate levels of action control, intervening only where necessary.[4]

Not all of us can engage in the finger acrobatics required for playing Chopin or Liszt. But many of us regularly engage in a similarly remarkable motor skill on another type of keyboard. I am writing this book on a laptop equipped with a standard QWERTY keyboard, named for the first six letters of the top row. The history of why the QWERTY keyboard, designed by politician and amateur inventor Christopher Latham Sholes in the 1860s, came into being is murky (the earliest typewriters instead had all twenty-six letters of the alphabet organized in a row from A to Z, which its inventors assumed would be the most efficient arrangement). One story is that it was to prevent the early typewriters from getting jammed. Another is that it helped telegraph operators, who received Morse code, quickly transcribe closely related letters in messages. And yet another is that Remington, the first major typewriter manufacturer, wanted to stick with QWERTY to ensure brand loyalty from typists who had trained on its proprietary system.

Whichever theory is correct, the English-speaking world's QWERTY typewriter has led millions of people to acquire a highly proficient but largely unconscious motor skill. If you are a regular computer user, close your eyes and try to imagine where the letters fall on your keyboard (with the exception of the letters Q-W-E-R-T-Y!). It is not easy, and if you are like me, can only really be done by pretending to type out a word. This neat dissociation between motor skill and conscious awareness makes typing a perfect test bed for studying the different kinds of algorithms involved in unconsciously monitoring and controlling our actions. Typing can also be studied with beautiful precision in the lab: the initiation and timing of keystrokes can be logged by a computer and the movements of people's fingers captured by high-resolution cameras.

Using these methods, the psychologists Gordon Logan and Matthew Crump have carried out detailed and creative experiments to probe how people type. In one of their experiments, people were asked to type out the answers to a classic psychological test, the Stroop task. In the Stroop, people are asked to respond to the color of the ink a word is written in—typing "blue" for blue ink and "red" for red ink, for instance. This is straightforward for most words, but when the words themselves are color words (such as the word "green" written in blue ink, "purple" written in red ink, and so on) it becomes much more difficult, and people slow down and make errors when the word and the ink color don't match. But despite being slower to initiate typing the word, they were no slower to type the letters *within* the word once they had gotten started (for instance, b-l-u-e). This led to the hypothesis that there are multiple action control loops at work: a higher-level loop governing the choice of which word to type, and a lower-level loop that takes this information and works out which keys need to be pressed in which order.[5]

Not only are there multiple levels of action control, but the higher levels know little about the workings of the lower levels. We know this because one of the easiest ways to screw up someone's typing is to ask them to type only the letters in a sentence that would normally be typed by the left (or right) hand. Try sitting at a keyboard and typing only the left-hand letters in the sentence "The cat on the mat" (on a QWERTY keyboard you should produce something like "Tecatteat," depending on whether you normally hit the space bar with your right or left thumb). It is a fiendishly difficult and frustrating task to assign letters to hands. And yet the lower-level loop controlling our keystrokes does this continuously, at up to seventy words per minute! Some part of us does know the correct hand, but it's not able to get the message out.[6]

Staying the Course

These experiments suggest that fine-scale unconscious adjustments are continuously being made to ensure that our actions stay on track. Occasionally, these unconscious monitoring processes become exposed, similar to how visual illusions revealed the workings of perceptual inference in the previous chapter. For instance, when I commute to work on the London Tube, I have to step onto a series of moving escalators, and I rely on my body making rapid postural adjustments to stop me from falling over when I do so. But this response is so well learned that if the escalator is broken and stationary, it's difficult to stop my motor system from automatically correcting for the impact of the usually moving stairs—so much so that I now have a higher-level expectation that I will stumble slightly going onto a stationary escalator.[7]

In a classic experiment designed to quantify this kind of rapid, automatic error correction, Pierre Fourneret and Marc Jeannerod asked volunteers to move a computer cursor to a target on a screen. By ensuring that participants' hands were hidden (so that they could see only the cursor), the researchers were able to introduce small deviations to the cursor position and observe what happened. They found that when the cursor was knocked off course, people immediately corrected it without being aware of having done so. Their paper concluded: "We found that subjects largely ignored the actual movements that their hand had performed." In other words, a low-level system unconsciously monitors how we are performing the task and corrects—as efficiently as possible—any deviations away from the goal.[8]

One part of the brain that is thought to be critical for supporting these adjustments is known as the cerebellum—from the Latin for "little brain." The cerebellum looks like a bolt-on secondary brain, sitting underneath the main brain. But it actually contains over 80 percent of your neurons, around sixty-nine billion of the eighty-five billion total. Its circuitry is a thing of regimented beauty, with millions of so-called parallel fibers crossing at right angles with another type of brain cell known as Purkinje neurons, which have huge, elaborate dendritic trees. Inputs come in from the cortex in a series of loops, with areas in the cortex projecting to the same cerebellar regions from which they receive input. One idea is that the cerebellum receives a copy of the motor command sent to the muscles, rather like receiving a carbon copy of an email. It then generates the expected sensory consequences of the action, such as the fact that my hand should be smoothly progressing toward the target. If this expectation does not match the sensory data about where my hand actually is, rapid adjustments can be made to get it back on course.[9]

In engineering, this kind of architecture is known as a forward model. Forward models first predict the consequences of a particular motor command and then track the discrepancy between the current state and what is expected to happen to provide small corrections as and when they are needed. Since I was a child I have loved to sail, either racing in smaller dinghies or cruising on bigger boats. When we are sailing from one place to the next, I can use a simple forward model to help counteract the effects of the tide on the position of the boat. If I plot a course to a harbor where I want to end up, the GPS tells me whether I'm left or right of the straight-line track, and I can correct accordingly without worrying about my overall heading. Often this results in the boat crabbing sideways into the tide, similar to how you would instinctively row upstream when crossing a river. To a casual observer, it looks like the adjustment for the tide was carefully planned in advance, when in fact it was the result of lots of small corrections based on local error signals.

In this kind of algorithm, what matters is keeping track of deviations from what I planned to happen. This means that information coming in from my senses can simply be ignored if it is in line with what I expect—another feature of predictive control that is borne out by experiments on human volunteers. For instance, if I were to passively move your arm, your brain would receive information that your arm is moving from the change in position of the joints and muscles. But if I move my arm myself, this sensory feedback is dampened, because it is exactly in line with what I expected to happen (this is also why you can't tickle yourself). These neural algorithms set up to detect deviations in our movements can lead to some counterintuitive phenomena. If you throw a punch in a boxing ring or a bar brawl, your unfortunate punch-recipient would feel the punch more acutely on his face than you, the punch-thrower, would on your hand. This is because the punch-thrower expects to feel the punch, whereas the punch-recipient does not. If the punch-recipient decides to retaliate, he will then throw a stronger punch than the punch-thrower believes he has thrown, in order to match the punch he feels he has just received. And so on, in a vicious cycle of escalating tit for tat. If you have ever sat in the front seat of a car while two kids are fighting in the back, you will know how this scenario can play out.[10]

What all these experiments tell us is that there are a range of self-monitoring processes churning away in the absence of conscious awareness. We are able to act smoothly and swiftly, usually without thinking about it, thanks to the power and flexibility of predictive control. When I step onto a moving escalator during rush hour, small corrections to my posture are made based on local algorithms, just as subtle adjustments to the course of a sailing boat are made to take into account effects of the tide. But if the deviation in what we expected to happen is large—if we are missing the target by a mile—then minor corrections or adjustments are unlikely to be able to bring it back into line. This is the point at which unconscious adjustments to our actions morph into conscious recognition of making an error.

From Detecting Errors to Learning About Ourselves

One of the first studies of how we become aware of our errors was carried out by psychologist Patrick Rabbitt in the 1960s. He designed a difficult, repetitive task involving pushing buttons in response to sequences of numbers. The actual task didn't matter too much—the clever part was that he also asked people to push another button if they detected themselves making an error. Rabbitt precisely measured the time it took for these additional button presses to occur, finding that people were able to correct their own errors very quickly. In fact, they realized they had made an error on average forty milliseconds faster than their fastest responses to external stimuli. This elegant and simple analysis proved that the brain was able to monitor and detect its own errors via an efficient, internal computation, one that did not depend on signals arriving from the outside world.

This rapid process of error detection can lead to an equally rapid process of error correction. In a simple decision about whether a stimulus belongs to category A or B, within only tens of milliseconds after the wrong button is pressed, the muscles controlling the correct response begin to contract in order to rectify the error. And if these corrective processes happen fast enough, they may prevent the error from occurring in the first place. For instance, by the time our muscles are contracting and we are pressing the send button on a rash email, we might have accumulated additional evidence to suggest that this is not a good idea and withhold the critical mouse click at the last moment.[11]

A couple of decades after Rabbitt's experiment, the brain processes that support internal error detection were beginning to be discovered. In his PhD thesis published in 1992, the psychologist William Gehring made electroencephalograph (EEG) recordings from participants while they performed difficult tasks. EEG uses a net of small electrodes to measure the changes in the electrical field outside the head caused by the combined activity of thousands of neurons inside the brain. Gehring found that a unique brain wave was triggered less than one hundred milliseconds after an error was committed. This rapid response helps explain why Rabbitt found that people were often able to very quickly recognize that they had made an error, even before they were told. This activity was labeled the error-related negativity (ERN), which psychologists now affectionately refer to as the "Oh shit!" response.[12]

We now know that the ERN occurs following errors on a multitude of tasks, from pressing buttons to reading aloud, and is generated by a brain region buried in the middle of the frontal lobe: the dorsal anterior cingulate cortex (dACC). This tell-tale neural signature of self-monitoring is already in place early in human development. In one experiment, twelve-month-old babies were flashed a series of images on a computer screen, and their eye movements recorded. Occasionally one of the images would be a face, and if the babies looked toward it, they would get a reward in the form of music and flashing colored lights. The occasions on which the baby failed to look at the face are errors in the context of the experiment—they did not perform the action that would get them the reward. On these occasions, EEG recordings showed a clear ERN, although somewhat delayed in time compared to what is typically seen in adults.[13]

We can think of the ERN as a special case of a "prediction error" signal. Prediction errors do exactly what they say on the tin—they keep track of errors in our predictions about the future, and they are a central feature of algorithms that can efficiently learn about the world. To see how prediction errors help us learn, imagine that a new coffee shop opens up near your office. You don't yet know how good it is, but they have taken care to buy a top-of-the-line espresso machine and get the ambience just right. Your expectations are high—you predict that the coffee will be good before you've even tasted it. When you sip your first cup, you find that it's not only good—it's one of the best cups of coffee you have had in a long time. The fact that the coffee was better than expected leads you to update your estimate, and it becomes your new favorite stop on the way in to work.

Now let's imagine a few weeks have gone by. The baristas have become complacent and the coffee is no longer as good as it used to be. It might still be good, but compared to what you expected, this is experienced as a negative error in your prediction, and you might feel a little more disappointed than usual.

The ability to make and update predictions depends on a famous brain chemical, dopamine. Dopamine is not only famous, but it is also commonly misunderstood and often referred to as the "pleasure" chemical in the popular media. It is true that dopamine is boosted by things that we enjoy, from money to food to sex. But the idea that dopamine simply signals the rewarding character of an experience is incorrect. In the 1990s, a now classic experiment was carried out by the neuroscientist Wolfram Schultz. He recorded signals from cells in the monkey midbrain that produce dopamine and deliver it to other brain areas. Schultz trained the monkeys to expect a drop of juice after a light was switched on in the room. Initially, the dopamine cells responded to the juice, consistent with the pleasure theory. But over time, the animals began to learn that the juice was always preceded by the light—they learned to expect the juice—and the dopamine response disappeared.[14]

An elegant explanation for the pattern of dopamine responses in these experiments is that they were tracking the error in the monkeys' prediction about the juice. Early on, the juice was unexpected—just like the unexpectedly good coffee from the new shop. But over time, the monkeys came to expect the juice every time they saw the light, just as we would come to expect good coffee every time we walked into the cafe. Around the same time that Schultz was performing his experiments, the computational neuroscientists Peter Dayan and Read Montague were building on classic work on trial-and-error learning in psychology. A prominent theory, the Rescorla-Wagner rule, proposed that learning should only occur when events are unexpected. This makes intuitive sense: If the coffee is just the same as yesterday, I don't need to alter my estimate of the goodness of the coffee shop. There is no learning to do. Dayan and Montague showed that versions of this algorithm provided an excellent match to the response of dopamine neurons. Shortly after Schultz, Dayan, and Montague's work was published, a series of studies by my former PhD adviser Ray Dolan discovered that the neural response in regions of the human brain that receive dopamine input closely tracks what one would expect of a prediction error signal. Together, these pioneering studies revealed that computing prediction errors and using them to update how we experience the world is a fundamental principle of how brains work.[15]

Now that we're armed with an understanding of prediction errors, we can begin to see how similar computations are important for self-monitoring. Occasionally we directly experience positive or negative feedback about our performance—on an assignment at school, for instance, or when we learn we have beaten our personal best over a half-marathon distance. But in many other areas of everyday life, the feedback may be more subtle, or even absent. One useful way of thinking about the ERN, then, is that it reflects an internal signal of reward or, more specifically, the absence of reward. It conveys the difference between what we expected (to perform well) and what actually happened (an error).

Consider sitting down to play a simple melody on the piano. Each note has a particular sound to it, but it would be strange to say that any note is "better" or "worse" than another. Played alone, an A is no more rewarding than a G-sharp. But in the context of a melody such as the opening of Grieg's Piano Concerto in A Minor, mistakenly playing a G-sharp instead of an A is immediately jarring, and we would wince at the clash. Even if no external feedback is involved, playing a wrong note is an error in how we expected to perform. And by keeping track of these performance errors, the brain can estimate whether it is doing well or badly—even in the absence of explicit feedback.[16]

By definition, errors are not usually made when we expect to make them—otherwise we would have been able to take steps to prevent them from happening. This feature of human error is used for comic effect in one of my favorite sketches from *The Fast Show*. A genial old man called Unlucky Alf turns to the camera and says in a broad northern English accent: "See that down there? They're digging a bloody great hole at the end

of the road. Knowing my luck I'll probably fall down that."
We watch and wait as he slowly ambles off down the road, at
which point an enormous gust of wind picks up and he is blown
into the hole. The sketch is funny because of the preparation
and foresight that went into unsuccessfully avoiding disaster. It
is more typical that we are surprised by errors precisely because
we didn't see them coming, and, like Homer in *The Simpsons*,
exclaim "D'oh!" when we recognize them after the fact.

A powerful way of implementing self-monitoring, then,
is to create predictions for how we expect to perform and
keep track of whether or not we performed as intended. If we
make a mistake, this is registered as a negative error in our
prediction of success. Remarkably, there is a beautiful symmetry between the brain circuitry involved in detecting external
rewards—such as whether the coffee was better or worse than
expected, or whether we received a recent bonus at work—and
those involved in tracking internal errors in our performance.
Both appear to rely on dopamine. For instance, in zebra finches,
introducing unexpected sounds into what the birds hear of their
own birdsong leads to reduced firing in dopamine neurons.
These particular dopamine neurons project to another brain
region involved in song learning, as if the dopamine is signaling
whether a recent effort was good or bad—sort of like a resident
talent-show judge in the bird's brain. The same circuits tracking
internal errors in song production also track external rewards,
just as we would expect if a common prediction error algorithm
is involved in learning about both the world and ourselves.[17]

Becoming Self-Aware

Let's recap. We have encountered cases in which error correction
is applied at multiple levels of the system—from detecting and
correcting changes in internal states (homeostasis) to ensuring
our actions remain in line with what we intended to do. Many

of these forms of self-monitoring are widespread in the animal kingdom and appear early in human development.[18] We have also seen that algorithms for estimating uncertainty and monitoring our internal states and actions are ubiquitous features of complex, self-regulating systems such as the human brain. These two building blocks form the core of what psychologists refer to as implicit metacognition—forms of self-monitoring that often proceed automatically and unconsciously. In contrast, explicit metacognition refers to those aspects of metacognition of which we are also consciously aware. When I become convinced that I have made a hash of a task at work, then I am engaging explicit metacognition.

A useful (if coarse) analogy for the relationship between implicit and explicit metacognition is the interaction between the pilots and autopilot of a modern airliner. The aircraft has an electronic "brain" in the form of its autopilot, which provides fine-grain self-monitoring of the plane's altitude, speed, and so on. The pilots, in turn, perceive and monitor the workings of the plane's autopilot, and such monitoring is governed by the workings of their (biological) brains. The interaction between pilot and autopilot is a rudimentary form of "aircraft awareness"—the pilots are tasked with being aware of what the plane's autopilot is doing and intervening where necessary. The same is true of the explicit and implicit aspects of metacognition, except now the interaction is all taking place within a single brain.

This does not mean there is the equivalent of an internal pilot sitting there monitoring what is happening in our heads. The concepts and models we use to describe how the mind works tend to be different from the concepts and models we use to describe its implementation in neural hardware. As an analogy, it makes sense to talk about the words of this book "existing" in my word-processing software, but it makes less sense to search for the words in the 1s and 0s zipping around my laptop's circuit board. In the same way, we might talk of self-awareness as involving something that "monitors" and "observes" other cognitive processes (a psychological or computational level of analysis), but this does not mean there is an actual monitor or observer to be found when we peer inside the head. The field of cognitive neuroscience is making us increasingly familiar with the notion that there is no single place in the brain where feelings happen or decisions are made, and the same is true of metacognition—there is no single location where self-awareness "happens."[19]

At the psychological level, though, the picture from cognitive science is that many of the relevant processes needed to "fly" the mind and body can be handled on autopilot, without involving explicit metacognition. A vast array of predictions and prediction errors are triggering continual adjustments to keep our mental planes flying straight and level—but for the most part they remain hidden from view, just as airline pilots are often oblivious to the continual adjustments an autopilot is making to keep their plane locked to a height of thirty thousand feet.

Many animals have this rich capacity for implicit metacognition—they can sense when they are uncertain and track errors in their actions. This helps them pass metacognitive tests such as Smith's uncertain-response test that we encountered in the previous chapter. Human infants as young as twelve months old also display sophisticated capacities for implicit metacognition. But by the time we reach adulthood, most of us have also acquired an explicit form of metacognition that allows us to consciously think about our own minds and those of others.[20]

The question that remains, then, is *why*? Why did we gain a remarkable ability to become aware of ourselves? Implicit metacognition—our vast array of unconscious autopilots—seems to be doing just fine without it. Why did evolution bother making any of this conscious?

3

KNOWING ME, KNOWING YOU

> Just as our picture of the physical world is a fantasy constrained by sensory signals, so our picture of the mental world, of our own or of others, is a fantasy constrained by sensory signals about what we, and they, are doing and saying.
>
> **—CHRIS FRITH,** *Making Up the Mind*

Unfortunately, we cannot go back in time and measure self-awareness in our now-extinct ancestors. But one elegant story about the origins of self-awareness goes like this: At some point in our evolutionary history, humans found it important to keep track of what other people were thinking, feeling, and doing. Psychologists refer to the skills needed to think through problems such as "Sheila knows that John knows that Janet doesn't know about the extra food" as theory of mind, or mindreading for short. After mindreading was established, the story goes, at some point it gradually dawned on our ancestors that they could apply such tools to think about themselves.

This particular transition in human history may have occurred somewhere between seventy thousand and fifty thousand years ago, when the human mind was undergoing substantial cognitive changes. Discoveries of artifacts show that the wearing of jewelry such as bracelets and beads became more common around this time—suggesting that people began to care about and understand how they were perceived by others. The first cave art can also be dated to a similar era, emerging in concert in locations ranging from Chauvet in France to Sulawesi in Indonesia. These haunting images show stencils of the hands of the artist or lifelike and beautiful drawings of animals such as bison or pigs. It's impossible to know for sure why they were created, but it is clear that these early human artists had some appreciation of how their paintings influenced other minds—whether the minds of other humans or the minds of the gods.[1]

The idea that there is a deep link between awareness of ourselves and awareness of others is often associated with the Oxford philosopher Gilbert Ryle. He argued that we self-reflect by applying the tools we use to understand other minds to ourselves: "The sorts of things that I can find out about myself are the same as the sorts of things that I can find out about other people, and the methods of finding them out are much the same."[2] Ryle's proposal is neatly summarized by an old *New Yorker* cartoon in which a husband says to his wife, "How should I know what I'm thinking? I'm not a mind reader."

I first encountered Ryle's ideas after finishing my PhD at UCL. I had just moved to New York to start my postdoctoral fellowship under the guidance of Nathaniel Daw, an expert on computational models of the brain. I had two main goals for this postdoc: the first was to teach myself some more mathematics (having dropped the subject much too early, at the age of sixteen), and the second was to think about ways of building computational models of self-awareness. I initially intended to visit NYU for a year or two, but New York is a difficult city to leave and I kept extending my stay. This was lucky for two reasons: it gave Nathaniel and me the chance to figure out our model of metacognition (which, like most science worth doing, took much longer than expected), and it gave me a chance to meet my wife, who was working as a diplomat at the United Nations.

The model Nathaniel and I developed is known as the second-order model of metacognition. The idea is that when we self-reflect, we use the same computational machinery as when we think about other people, just with different inputs. If a second-order Rylean view of self-awareness is on the right track, then we should be able to discover things about how self-awareness works from studying social cognition. More specifically—how we think about other minds.[3]

Thinking About Minds

A defining feature of mindreading is that it is recursive—as in, "Keith believes that Karen thinks he wants her to buy the cinema tickets." Each step in this recursion may be at odds with reality. For instance, Karen may not be thinking that at all, and even if she is, Keith may not want her to buy the tickets. The success or failure of mindreading often turns on this ability to represent the possibility that another person's outlook on a situation might be at odds with our own. Sometimes this can be difficult to do, as captured by another *New Yorker* cartoon caption: "Of course I care about how you imagined I thought you perceived I wanted you to feel."

Mismatches between what we think others are thinking and what they are actually thinking can be rich sources of comic misunderstanding. When US president Jimmy Carter gave a speech at a college in Japan in 1981, he was curious to know how the interpreter translated his joke, as it sounded much shorter than it should have been and the audience laughed much harder than they usually did back home. After much cajoling, the interpreter finally admitted that he had simply said, "President Carter told a funny story. Everyone must laugh."

As adults, we usually find mindreading effortless; we don't have to churn through recursive calculations of who knows what. Instead, in regular conversation, we have a range of shared assumptions about what is going on in each other's minds—what we know, and what others know. I can text my wife and say, "I'm on my way," and she will know that by this I mean I'm on my way to collect our son from day care, not on my way home, to the zoo, or to Mars. But this fluency with reading minds is not something we are born with, nor is it something that is guaranteed to emerge in development.

In a classic experiment designed to test the ability for mindreading, children were told stories such as the following: Maxi has put his chocolate in the cupboard. While Maxi is away, his mother moves the chocolate from the cupboard to the drawer. When Maxi comes back, where will he look for the chocolate? Only if children are able to represent the fact that Maxi thinks the chocolate is still in the cupboard (known as a false belief, because it conflicts with reality) can they answer correctly. Until the age of four, children often fail this test, saying that Maxi will look for the chocolate where it actually is, rather than where he thinks it is. Children with autism tend to struggle with this kind of false-belief test, suggesting that they have problems with smoothly tracking the mental states of other people. Problems with mindreading in autism can be quite specific. In one experiment, autistic children were just as good at, if not better than, children of a similar age when asked to order pictures that implied a physical sequence (such as a rock rolling down a hill) but selectively impaired at ordering pictures that need an understanding of changes in mental states (such as a girl being surprised that someone has moved her teddy bear).[4]

A similar developmental profile is found in tests of children's self-awareness. In one experiment from Simona Ghetti's lab, three-, four-, and five-year-old children were first asked to memorize a sequence of drawings of objects, such as a boat, baby carriage, and broom. They were then asked to pick, from pairs of drawings, which one they had seen before. After each pair, the children were asked to indicate their confidence by choosing a picture of another child who most matched how they were feeling: very unsure, a little unsure, or certain they were right. Each age group had a similar level of memory performance—they all tended to forget the same number of pictures. But their metacognition was strikingly different. Three-year-olds' confidence ratings showed little difference between correct and incorrect decisions. Their ability to know whether they were right or wrong was poor. In contrast, the four- and five-year-olds showed good metacognition, and were also more likely to put high-confidence answers forward to get a prize. Just as adults taking a multiple-choice exam may elect to skip a question when they feel uncertain, by the time children reach four years old they are able to notice when they might be wrong and judiciously put those answers to one side.[5]

It is striking, then, that the ability to estimate whether someone else has a different view of the world—mindreading—emerges in children around the same time that they acquire explicit metacognition. Both of these abilities depend on being able to hold reality at arm's length and recognizing when what we believe might be deviating from reality. In other words, to understand that we ourselves might be incorrect in our judgments about the world, we engage the same machinery used for recognizing other people's false beliefs. To test this idea, in one study children were first presented with "trick" objects: a rock that turned out to be a sponge, or a box of Smarties that actually contained pencils. When asked what they thought the object was when they first perceived it, three-year-olds said that they knew all along that the rock was a sponge and that the Smarties box was full of pencils. But by the age of five, most children recognized that their first impression of the object was false—they successfully engaged in self-doubt. In another version of this experiment, children sat opposite each other and various boxes were placed on a table in between them. Each box contained a surprise object, such as a coin or a piece of chocolate, and one of the children was told what was in the box. Only half of three-year-olds correctly realized that they didn't know what was in the box when they had not been told, whereas by the age of five all of the children were aware of their own ignorance.[6]

It is possible, of course, that all this is just a coincidence—that metacognition and mindreading are two distinct abilities that happen to develop at a similar rate. Or it could be that they are tightly intertwined in a virtuous cycle, with good meta-cognition facilitating better mindreading and vice versa. One way to test this hypothesis is to ask whether differences in min-dreading ability early in childhood statistically predict later self-awareness. In one study, at least, this was found to be the case: mindreading ability at age four predicted later self-awareness, even when controlling for differences in language development. Another way to test this hypothesis is to ask whether the two abilities—metacognition and mindreading—interfere with each other, which would indicate that they rely on a common mental resource. Recent data is consistent with this prediction: thinking about what someone else is feeling disrupts the ability to self-reflect on our own task performance but does not affect other aspects of task performance or confidence. This is exactly what we would expect if awareness of ourselves and others depends on common neural machinery.[7]

Self-awareness does not just pop out of nowhere, though. As we saw in the previous chapter, laboratory experiments show that the building blocks of self-monitoring are already in place in infants as young as twelve months old. By monitoring eye movements, it is possible to detect telltale signs that toddlers under the age of three are also sensitive to false beliefs. From around two years of age, children begin to evaluate their behavior against a set of standards or rules set by parents and teachers, and they show self-conscious emotions such as guilt and embarrassment when they fall short and pride when they succeed. This connection between metacognition and self-conscious emotions was anticipated by Darwin, who pointed out that "thinking what others think of us . . . excites a blush."[8]

An understanding of mirrors and fluency with language are also likely to contribute to the emergence of self-awareness in childhood. In the famous mirror test, a mark is made on the test subject's body, and if they make movements to try to rub it off, this is evidence that they recognize the person in the mirror as themselves, rather than someone else. Children tend to pass this test by the age of two, suggesting that they have begun to understand the existence of their own bodies. Being able to recognize themselves in a mirror also predicts how often children use the personal pronoun (saying things like "I," "me," "my," or "mine"), suggesting that awareness of our bodies is an important precursor to becoming more generally self-aware.[9]

Linguistic fluency also acts as a booster for recursive thought. The mental acrobatics required for both metacognition and mindreading share the same set of linguistic tools: thinking "I believe x" or "She believes x." Again, it is likely to be no coincidence that the words we use to talk about mental states—such as "believe," "think," "forget," and "remember"—arise later in childhood than words for bodily states ("hungry!"), and that this developmental shift occurs in English, French, and German around the same time as children acquire an understanding of other minds. Just like language, mindreading is recursive, involving inserting what you believe to be someone else's state of mind into your own.[10]

Watching self-awareness emerge in your own child or grandchild can be a magical experience. My son Finn was born halfway through writing this book, and around the time I was finalizing the proofs, when he was eighteen months old, we moved to a new apartment with a full-length mirror in the hallway. One afternoon, when we were getting ready to go out to the park, I quietly watched as he began testing his reflection in the mirror, gradually moving his head from side to side. He then slowly put one hand into his mouth while watching his reflection (a classic example of "mirror touch") and a smile broke out across his face as he turned to giggle at me.

There may even be a connection between the first glimmers of self-awareness and the playfulness of childhood. Initial evidence suggests that markers of self-awareness (such as mirror-self-recognition and pronoun use) are associated with whether or not children engage in pretend play, such as using a banana as a telephone or creating an elaborate tea party for their teddy bears. It's possible that the emergence of metacognition allows children to recognize the difference between beliefs and reality and create an imaginary world for themselves. There is a lovely connection here between the emergence of play in children and the broader role of metacognition and mindreading in our appreciation of theater and novels as adults. We never stop pretending—it is just the focus of our pretense changes.[11]

Some, but not all, of these precursor states of self-awareness that we find in children can also be identified in other animals. Chimpanzees, dolphins, and an elephant at the Bronx Zoo in New York have been shown to pass the mirror test. Chimpanzees can also track what others see or do not see. For instance, they know that when someone is blindfolded, they cannot see the food. Dogs also have similarly sophisticated perspective-taking abilities: stealing food when a human experimenter is not looking, for instance, or choosing toys that are in their owner's line of sight when asked to fetch. But only humans seem to be able to understand that another mind may hold a genuinely different view of the world to our own. For instance, if chimpanzee A sees a tasty snack being moved when chimpanzee B's back is turned (the ape equivalent of the Maxi test), chimpanzee A does not seem to be able to use this information to their advantage to sneakily grab the food. This is indeed a harder computational problem to solve. To recognize that a belief might be false, we must juggle two distinct models of the world. Somehow, the human brain has figured out how to do this—and, in the process, gained an unusual ability to think about itself. In the remainder of this chapter, we are going to explore how the biology of the human brain enables this remarkable conjuring trick.[12]

Machinery for Self-Reflection

A variety of preserved brains can be seen at the Hunterian Museum in London, near the law courts of Lincoln's Inn Fields. The museum is home to a marvelous collection of anatomical specimens amassed by John Hunter, a Scottish surgeon and scientist working at the height of the eighteenth-century Enlightenment. I first visited the Hunterian Museum shortly after starting my PhD in neuroscience at UCL. I was, of course, particularly interested in the brains of all kinds—human and animal—carefully preserved in made-to-measure jars and displayed in rooms surrounding an elegant spiral staircase. All these brains had helped their owners take in their surroundings, seek out food, and (if they were lucky) find themselves a mate. Before they were each immortalized in formaldehyde, these brains' intricate networks of neurons fired electrical impulses to ensure their host lived to fight another day.

Each time I visited the Hunterian, I had an eerie feeling when looking at the human brains. On one level, I knew that they were just like any of the other animal brains on display: finely tuned information-processing devices. But it was hard to shake an almost religious feeling of reverence as I peered at the jars. *Each and every one of them once knew that they were alive.* What is it about the human brain that gives us these extra layers of recursion and allows us to begin to know ourselves? What is the magic ingredient? *Is* there a magic ingredient?

One clue comes from comparing the brains of humans and other animals. It is commonly assumed that humans have particularly large brains for our body size—and this is partly true, but not in the way that you might think. In fact, comparing brain and body size does not tell us much. It would be like concluding that a chip fitted in a laptop computer is more powerful than the same chip fitted into a desktop, just because the laptop has a smaller "body." This kind of comparison does not tell us much about whether the brains of different species—our own included—are similar or different.

Instead, the key to properly comparing the brains of different species lies in estimating the number of neurons in what neuroscientist Suzana Herculano-Houzel refers to as "brain soup." By sticking the (dead!) brains of lots of different species in a blender, it is possible to plot the actual number of cells in a brain against the brain mass, enabling meaningful comparisons to be made.

After several painstaking studies of brains of all shapes and sizes, a fascinating pattern has begun to emerge. The number of neurons in primate brains (which include monkeys, apes such as chimpanzees, and humans) increases linearly with brain mass. If one monkey brain is twice as large as another, we can expect it to have twice as many neurons. But in rodents (such as rats and mice), the number of neurons increases more slowly and then begins to flatten off, in a relationship known as a power law. This means that to get a rodent brain with ten times the number of neurons, you need to make it forty times larger in mass. Rodents are much less efficient than primates at packing neurons into a given brain volume.[13]

It's important to put this result in the context of what we know about human evolution. Evolution is a process of branching, rather than a one-way progression from worse to better. We can think of evolution like a tree—we share with other animals a common ancestor toward the roots, but other groups of species branched off the trunk many millions of years ago and then continued to sprout subbranches, and subbranches of sub-branches, and so on. This means that humans (*Homo sapiens*) are not at the "top" of the evolutionary tree—there is no top to speak of—and instead we just occupy one particular branch. It is all the more remarkable, therefore, that the same type of neuronal scaling law seen in rodents is found both in a group that diverged from the primate lineage around 105 million years ago (the afrotherians, which include the African elephant) and a group that diverged much more recently (the artiodactyls, which include pigs and giraffes). Regardless of their position on the tree, it seems that primates are evolutionary outliers—but, relative to other primates, humans are not.[14]

What seems to pick primates out from the crowd is that they have unusually efficient ways of cramming more neurons into a given brain volume. In other words, although a cow and a chimpanzee might have brains of similar weight, we can expect the chimpanzee to have around twice the number of neurons. And, as our species is the proud owner of the biggest primate brain by mass, this creates an advantage when it comes to sheer number of neurons. The upshot is that what makes our brains special is that (a) we are primates, and (b) we have big heads![15]

We do not yet know what this means. But, very roughly, it is likely that there is simply more processing power devoted to so-called higher-order functions—those that, like self-awareness, go above and beyond the maintenance of critical functions like homeostasis, perception, and action. We now know that there are large swaths of cortex in the human brain that are not easy to define as being sensory or motor, and are instead tradition-ally labeled as association cortex—a somewhat vague term that refers to the idea that these regions help associate or link up many different inputs and outputs.

Regardless of the terminology we favor, what is clear is that association cortex is particularly well-developed in the human brain compared to other primates. For instance, if you examined different parts of the human prefrontal cortex (which is part of the association cortex, located toward the front of the brain) under the microscope, you would sometimes find an extra layer of brain cells in the ribbonlike sheet of cortex known as a granular layer. We still don't fully understand what this additional cell layer is doing, but it provides a useful anatomical landmark with which to compare the brains of different species. The granular portion of the PFC is considerably more folded and enlarged in humans compared to monkeys and does not exist at all in rodents. It is these regions of the association cortex—particularly the PFC—that seem particularly important for human self-awareness.[16]

Many of the experiments that we run in our laboratory are aimed at understanding how these parts of the human brain support self-awareness. If you were to volunteer at the Wellcome Centre for Human Neuroimaging, we would meet you in our colorful reception, decorated with images of different types of scanners at work, and then we would descend to the basement where we have an array of large brain scanners, each arranged in different rooms. After filling in forms to ensure that you are safe to enter the scanning suite—magnetic resonance imaging (MRI) uses strong magnetic fields, so volunteers must have no metal on them—you would hop onto a scanner bed and see various instructions on a projector screen above your head. While the scanner whirs away, we would ask you a series of questions: Do you remember seeing this word? Which image do you think is brighter? Occasionally, we might also ask you to reflect on your decisions: How confident are you that you got the answer right?

MRI works by using strong magnetic fields and pulses of radio waves to pinpoint the location and type of tissue in the body. We can use one type of scan to create high-resolution three-dimensional pictures of the brains of volunteers in our experiments. By tweaking the settings of the scanner, rapid snapshots can also be taken every few seconds that track changes in blood oxygen levels in different parts of the brain (this is known as functional MRI, or fMRI). Because more vigorous neural firing uses up more oxygen, these changes in blood oxygen levels are useful markers of neural activity. The fMRI signal is very slow compared to the rapid firing of neurons, but, by applying statistical models to the signal, it is possible to reconstruct maps that highlight brain regions as being more or less active when people are doing particular tasks.

If I put you in an fMRI scanner and asked you to think about yourself, it's a safe bet that I would observe changes in activation in two key parts of the association cortex: the medial PFC and the medial parietal cortex (also known as the precuneus), which collectively are sometimes referred to as the cortical midline structures. These are shown in the image on page 68, which was created from software that searches the literature for brain activation patterns that are consistent with a particular search term, which in this case was "self-referential." Robust activation of the medial PFC is seen in experiments where people are asked to judge whether adjectives such as "kind" or "anxious" apply to either themselves or someone famous, such as the British queen. Retrieving memories about ourselves, such as imagining the last time we had a birthday party, also activates the same regions. Remarkably, and consistent with Ryle's ideas of a common system supporting mindreading and self-awareness, the same brain regions are also engaged when we are thinking about other people. How close these activity patterns match depends on how similar the other person is to ourselves.[17]

Brain imaging is a powerful tool, but it relies on correlating what someone is doing or thinking in the scanner with their

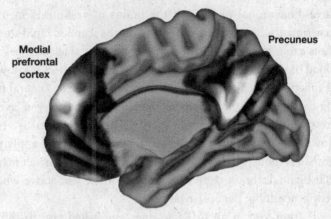

Medial surface activations obtained using the meta-analysis tool NeuroQuery in relation to the term "self-referential."

(https://neuroquery.org, accessed September 2020.)

pattern of neural activity. It cannot tell us whether a particular region or activity pattern is causally involved in a particular cognitive process. Instead, to probe causality, we can use stimulation techniques such as transcranial magnetic stimulation (TMS), which uses strong magnetic pulses to temporarily disrupt normal neural activity in a particular region of cortex. When TMS is applied to the parietal midline, it selectively affects how quickly people can identify an adjective as being relevant to themselves, suggesting that the normal brain processes in this region are important for self-reflection.[18]

Damage to these networks can lead to isolated changes in self-awareness—we may literally lose the ability to know ourselves. The first hints that brain damage could lead to problems with metacognition came in the mid-1980s. Arthur Shimamura, then a postdoctoral researcher at the University of California, San Diego, was following up on the famous discovery of patient "HM," who had become forever unable to form new memories after brain surgery originally carried out to cure his epilepsy.

The surgery removed HM's medial temporal lobe, an area of the brain containing the hippocampus and a region crucial for memory. Shimamura's patients, like HM, had damage to the temporal lobe, and therefore it was unsurprising that many of them were also amnesic. What was surprising was that some of his patients were also *unaware* of having memory problems. In laboratory tests, they showed a striking deficit in metacognition: they were unable to rate how confident they were in getting the answers right or wrong.

The subgroup of patients who showed this deficit in metacognition turned out to have Korsakoff's syndrome, a condition linked to excessive alcohol use. Korsakoff's patients often have damage not only to structures involved in memory storage, such as the temporal lobe, but also to the frontal lobe of the brain that encompasses the PFC. Shimamura's study was the first to indicate that the PFC is also important for metacognition.[19]

However, there was one concern about this striking result. All of Shimamura's patients were amnesic, so perhaps their metacognitive deficit was somehow secondary to their memory problems. This illustrates a general concern we should keep in mind when interpreting scientific studies of self-awareness. If one group appears to have poorer metacognition than another, this is less interesting if they also show worse perception, memory, or decision-making, for instance. Their loss of metacognition, while real, may be a consequence of changes in other cognitive processes. But if we still find differences in metacognition when other aspects of task performance are well matched between groups or individuals, we can be more confident that we have isolated a change in self-awareness that cannot be explained by other factors.

To control for this potential confounding variable, Shimamura needed to find patients with impaired metacognition but intact memory. In a second paper published in 1989, he and his colleagues reported exactly this result. In a group of patients who had suffered damage to their PFC, memory was relatively intact but metacognition was impaired. It seemed that damage to one set of brain regions (such as the medial temporal lobes) could lead to memory deficits but leave metacognition intact, whereas damage to another set of regions (the frontal lobes) was associated with impaired metacognition but relatively intact memory. This is known as a double dissociation and is a rare finding in neuroscience. It elegantly demonstrates that self-awareness relies on distinct brain processes that may be selectively affected by damage or disease.[20]

Shimamura's findings also helped shed light on a puzzling observation made around the same time by Thomas Nelson, one of the pioneers of metacognition research. Nelson was a keen mountaineer and combined his interests in climbing and psychology by testing his friends' memory as they were ascending Mount Everest. While the extreme altitude did not affect the climbers' ability to complete basic memory tests, it did affect their metacognition—they were less accurate at predicting whether they would know the answer or not. The oxygen content of the atmosphere at the summit of Everest (8,848 meters, or 29,029 feet) is around one-third of that at sea level. Because a lack of oxygen is particularly damaging to the functions of the PFC, this may explain why the climbers temporarily showed similar characteristics to Shimamura's patients.[21]

A few years later, the advent of functional brain imaging technology allowed this hypothesis to be directly tested by measuring the healthy brain at work. Yun-Ching Kao and her colleagues used fMRI to visualize changes in brain activity while volunteers were asked to remember a series of pictures such as a mountain scene or a room in a house. After viewing each picture, they were asked a simple question to tap into their metacognition: Will you remember this later on? Kao sorted people's brain activations into categories based on whether people actually remembered the picture and whether they *predicted* they would remember the picture. Being able to remember the picture was associated with increased activity in the temporal lobe, as expected. But the temporal lobe did not track people's metacognition. Instead, metacognitive judgments were linked to activation in the medial PFC—activity here was higher when people predicted they would remember something, regardless of whether they actually did. The activation of this region was strongest in people with better metacognition. Imaging of the healthy brain supports the conclusions from Shimamura's amnesic patients: that the PFC is a crucial hub for self-awareness.[22]

By the time we reach adulthood, most of us are adept at reflecting on what we know and what others know. Recently, Anthony Vaccaro and I surveyed the accumulating literature on mindreading and metacognition and created a brain map that aggregated the patterns of activations reported across multiple papers. In general, metacognition tended to engage regions that were more dorsal (above) and posterior (behind) the mindreading network. However, clear overlap between brain activations involved in metacognition and mindreading was observed in the ventral and anterior medial PFC. Thoughts about ourselves and others indeed seem to engage similar neural machinery, in line with a Rylean, second-order view of how we become self-aware.[23]

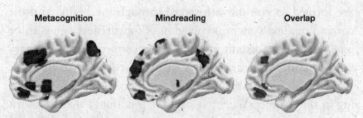

Brain activations obtained in a meta-analysis of metacognition compared to brain activations related to the term "mentalizing," from Neurosynth.

(Reproduced with permission from Vaccaro and Fleming, 2018.)

Breakthrough Powers of Recursion

We have already seen that animals share a range of precursors for self-awareness. The science of metacognition deals in shades of gray, rather than black or white. Other animals have the capacity for at least implicit metacognition: they can track confidence in their memories and decisions, and use these estimates to guide future behavior. It makes sense, then, that we can also identify neural correlates of confidence and metacognition in animal brains—for instance, patterns of neural activity in the frontal and parietal cortex of rodents and monkeys.[24]

Again, then, we have a picture in which self-awareness is a continuum, rather than an all-or-nothing phenomenon. Many of these precursors of self-awareness are seen in human infants. But it is also likely that adult humans have an unusual degree of self-awareness, all thanks to a sheer expansion of neural real estate in the association cortex, which, together with our fluency with language, provides a computational platform for building deep, recursive models of ourselves.[25]

We already encountered the idea that sensory and motor cortices are organized in a hierarchy, with some parts of the system closer to the input and others further up the processing chain. It is now thought that the association cortex also has a quasi-hierarchical organization to it. For instance, in the PFC, there is a gradient in which increasingly abstract associations are formed as you move forward through the brain. And the cortical midline system involved in self-awareness seems to be one of the most distant in terms of its connections to primary sensory and motor areas. It is likely no coincidence that this network is also reliably activated when people are quietly resting in the scanner. When we are doing nothing, we often turn our thoughts upon ourselves, trawling through our pasts and imagining our potential futures. The psychologist Endel Tulving referred to this aspect of metacognition as "autonoetic"—an awareness of ourselves as existing in memories from our past, in perceptions of the present, and in projections into the future.[26]

A link between metacognition and mindreading provides hints about the evolutionary driving forces behind humans acquiring a remarkable ability for self-awareness. Of course, much of this is speculation, as it is difficult to know how our mental lives were shaped by the distant past. But we can make some educated guesses. A rapid cortical expansion, thanks to the primate scaling rules, allowed humans to achieve an unprecedented number of cortical neurons. This was put to use in creating an ever more differentiated PFC and machinery that goes beyond the standard perception-action loop. But as Herculano-Houzel has pointed out, the human brain could not have expanded as it did without a radical increase in the number of calories. The fuel for this expansion may have come from a virtuous cycle of social cooperation, allowing more sophisticated hunting and cooking, fueling further cortical expansion, which allowed even greater cooperation and even greater caloric gain. This positive feedback loop is likely to have prized an ability to coordinate and collaborate with others. We have already seen that metacognition provides a unique benefit in social situations, allowing us to share what is currently on our minds and pool our perceptual and cognitive resources. In turn, mindreading becomes important to convert simple, one-way utterances into a joint understanding of what others are thinking and feeling. Many other animals have an ability for self-monitoring. But only humans have the ability (and the need) to explicitly represent the contents of the minds of themselves and others.[27]

Let's recap our journey so far. We have seen how simple systems can estimate uncertainty and engage in self-monitoring. Many of these building blocks for metacognition can operate unconsciously, providing a suite of neural autopilots that are widely shared across the animal kingdom and present early in human development. Self-awareness continues to crystallize in toddlers, becoming fully formed between the ages of three and four. But the emergence of self-awareness around the age of three is only the beginning of a lifetime of reflective thought. In the next chapter, we will see how a multitude of factors continue to buffet and shape the capacity for self-awareness throughout our adult lives. By harnessing these factors, we will also discover tools for deliberately boosting and shaping our powers of reflection.

4

BILLIONS OF SELF-AWARE BRAINS

> The biggest danger, that of losing oneself, can pass off
> in the world as quietly as if it were nothing; every other
> loss, an arm, a leg, five dollars, a wife, etc., is bound to
> be noticed.
>
> —SØREN KIERKEGAARD, *The Sickness unto Death*

On February 12, 2002, then US secretary of defense Donald
Rumsfeld was asked a question by NBC correspondent Jim
Miklaszewski about evidence that the Iraqi government had
weapons of mass destruction. Rumsfeld's response was to become famous:

> As we know, there are known knowns; there are things we
> know we know. We also know there are known unknowns;
> that is to say we know there are some things we do not
> know. But there are also unknown unknowns—the ones
> we don't know we don't know. And if one looks through-
> out the history of our country and other free countries, it is
> the latter category that tend to be the difficult ones.

The idea of known and unknown unknowns is usually applied to judgments about the external world (such as weapons or economic risks). Rumsfeld's argument was influential in persuading the United States to invade Iraq, after both the White House and UK government agreed it was too dangerous to do nothing about one particular unknown unknown: the ultimately illusory weapons of mass destruction. But we can also apply the same categories to judgments about ourselves, providing a tool for quantifying self-awareness.

This is a strange idea the first time you encounter it. A useful analogy is the index of a book. Each of the entries in the index usually points to the page number containing that topic. We can think of the index as representing the book's knowledge about itself. Usually the book's metacognition is accurate—the entries in the index match the relevant page number containing that topic. But if the index maker has made a mistake and added an extra, irrelevant entry, the book's metacognition will be wrong: the index will "think" that it has pages on a topic when it actually doesn't. Similarly, if the index maker has omitted a relevant topic, the book will have information about something that the index does not "know" about—another form of inaccurate metacognition.

In a similar way to the index of a book, metacognitive mechanisms in the human mind give us a sense of what we do and don't know. There are some things we know, and we know that we know them (the index matches the book), such as an actor's belief that he'll be able to remember his lines. There are other things we know that we don't know, or won't be able to know, such as that we're likely to forget more than a handful of items on a shopping list unless we write them down. And just like the unknown unknowns in Rumsfeld's taxonomy, there are also plenty of cases in which we don't know that we don't know— cases in which our self-awareness breaks down.

The measurement and quantification of self-awareness has a checkered history in psychology, although some of the field's initial pioneers were fascinated by the topic. In the 1880s, Wilhelm Wundt began collecting systematic data on what people thought about their perceptions and feelings, spending thousands of hours in the lab painstakingly recording people's judgments. But, partly because the tools to analyze this data had not yet been invented, the resulting research papers on what became known as introspectionism were criticized for being unreliable and not as precise as other branches of science. This led to a schism among the early psychologists. In one camp sat the behaviorists, who argued that self-awareness was irrelevant (think rats in mazes). In the other camp sat the followers of Freud, who believed in the importance of self-awareness but thought it better investigated through a process of psychoanalysis rather than laboratory experiments.[1]

Both sides were right in some ways and wrong in others. The behaviorists were right that psychology needed rigorous experiments, but they were wrong that people's self-awareness in these experiments did not matter. The Freudians were right to treat self-awareness as important and something that could be shaped and changed, but wrong to ground their approach in storytelling rather than scientific experiments. Paradoxically, to create a science of self-awareness, we cannot rely only on what people tell us. By definition, if you have poor metacognition, you are unlikely to know about it. Instead, a quantitative approach is needed.[2]

One of the first such attempts to quantify the accuracy of metacognition was made in the 1960s by a young graduate student at Stanford named Joseph Hart. Hart realized that people often think they know more than they can currently recall, and this discrepancy provides a unique window onto metacognition. For instance, if I ask you, "What is Elton John's real name?"

you might have a strong feeling that you know the answer, even if you can't remember it. Psychologists refer to these feelings as "tip of the tongue" states, as the answer feels as though it's just out of reach. Hart found that the strength of these feelings in response to a set of quiz questions predicted whether people would be able to subsequently recognize the correct answers. In other words, people have an accurate sense of knowing that they know, even if they cannot recall the answer.[3]

Hart's approach held the key to developing quantitative measures of self-awareness. His data showed it was possible to collect systematic data on people's judgments about themselves and then compare these judgments to the reality of their cognitive performance. For instance, we can ask people questions such as:

- Will you be able to learn this topic?

- How confident are you about making the right decision?

- Did you really speak to your wife last night or were you dreaming?

All these questions require judging the success of another cognitive process (specifically, learning, decision-making, and memory). In each case, it is then possible to assess if our judgments track our performance: we can ask whether judgments of learning relate to actual learning performance, whether judgments of decision confidence relate to the likelihood of making good decisions, and so on. If we were able to observe your confidence over multiple different occasions and record whether your answers were actually right or wrong, we could build up a detailed statistical picture of the accuracy of your metacognition. We could summarize your responses using the following table:

	Confident	Less confident
Right	A	B
Wrong	C	D

The relative proportion of judgments that fall into each box in the table acts as a ruler by which we can quantify the accuracy of your metacognition. People with better metacognition will tend to rate higher confidence when they're correct (box A) and lower confidence when they make errors (box D). In contrast, someone who has poorer metacognition may sometimes feel confident when they're actually wrong (box C) or not know when they're likely to be right (box B). The more As and Ds you have, and the fewer Bs and Cs, the better your metacognition—what we refer to as having good metacognitive sensitivity. Metacognitive sensitivity is subtly but importantly different from metacognitive bias, which is the overall tendency to be more or less confident. While on average I might be overconfident, if I am still aware of each time I make an error (the Ds in the table), then I can still achieve a high level of metacognitive sensitivity. We can quantify people's metacognitive sensitivity by fitting parameters from statistical models to people's confidence ratings (with names such as meta-d' and Φ). Ever more sophisticated models are being developed, but they ultimately all boil down to quantifying the extent to which our self-evaluations track whether we are actually right or wrong.[4]

What Makes One Person's Metacognition Better than Another's?

When I was starting my PhD in cognitive neuroscience at UCL in 2006, brain imaging studies such as those we encountered in the previous chapter were beginning to provide clear hints about the neural basis of self-awareness. What was lacking, however, were the tools needed to precisely quantify metacognition in the lab. In the first part of my PhD, I dabbled in developing these tools as a side project, while spending the majority of my time learning how to run and analyze brain imaging experiments. It wasn't until my final year that a chance discussion made me realize that we could combine neuroscience with this toolkit for studying self-awareness.

On a sunny July day in 2008, I had lunch in Queen Square with Rimona Weil, a neurologist working on her PhD in Geraint Rees's group at our Centre. She told me how she and Geraint were interested in what made individuals different from one another, and whether such differences were related to measurable differences in brain structure and function. In return, I mentioned my side project on metacognition—and almost simultaneously we realized we could join forces and ask what it is about the brain that made one person's metacognition better than another's. I was unaware then that this one project would shape the next decade of my life.

Rimona and I set out to relate individual differences in people's metacognition to subtle differences in their brain structure, as a first attempt to zero in on brain networks that might be responsible for metacognitive ability. In our experiment, people came into the laboratory for two different tests. First, they sat in a quiet room and performed a series of difficult judgments about visual images. Over an hour of testing, people made hundreds of decisions as to whether the first or second flashed image on a computer screen contained a slightly brighter patch. After every decision, they indicated their confidence on a six-point scale. If people made a lot of mistakes, the computer automatically made the task a bit easier. If they were doing well, the computer made the task a bit harder. This ensured everyone performed at a similar level and allowed us to focus on measuring their metacognition—how well they could track moment-to-moment changes in their performance.

This gave us two scores for each person: how adept they were at making visual discriminations and how good their metacognition was. Despite people performing equally well on the primary task, we still observed plenty of variation in their metacognitive sensitivity.

In a second part of the experiment, the same people came back to the lab and had their brains scanned using MRI. We collected two types of data: The first scan allowed us to quantify differences in the volume of gray matter (the cell bodies of neurons) in different regions of the brain. The second scan examined differences in the integrity of white matter (the connections between brain regions). Given the findings in other studies of patients with brain damage, we hypothesized that we would find differences in the PFC related to metacognition. But we didn't have any idea of what these differences might be.

The results were striking. People with better metacognition tended to have more gray matter in the frontal pole (also known as the anterior prefrontal or frontopolar cortex)—a region of the PFC toward the very front of the brain. They also had greater white matter integrity in bundles of fibers projecting to and from this region. Together, these findings suggest that we may have stumbled upon part of a brain circuit playing a role in supporting accurate self-awareness.[5]

This data is difficult to collect—involving many hours of painstaking psychophysics and brain scanning—but once it is assembled the analysis is surprisingly swift. One of the most widely used (and freely available) software packages for analyzing brain imaging data, called SPM, has been developed at our Centre. No one had previously looked for differences in the healthy brain related to metacognition, and there was a good

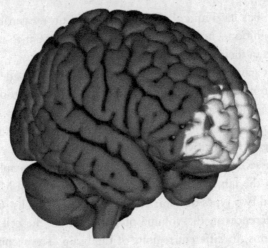

The frontal pole, part of the prefrontal cortex

chance of finding nothing. But after writing hundreds of lines of computer code to process the data, all that was needed was a single mouse click in SPM to see whether we had any results at all. It was a thrilling feeling when statistical maps, rather than blank brains, started to emerge on the screen.

Our research was exploratory, and voxel-based morphometry, as this technique is known, is a coarse and indirect measure of brain structure. In hindsight, we now know that this sample size was probably too small, or underpowered, for this kind of experiment. Statistical power refers to whether an experiment is likely to find an effect, given that it actually exists. The number of samples you need is governed by the size of the effect you expect to find. For instance, to determine whether men are statistically taller than women, I would perhaps need to sample fifteen to twenty of each to gain confidence in the difference and iron out any noise in my samples. But to establish that children are smaller than adults of either gender (a bigger effect size),

I would need to sample fewer of each category. In the case of brain imaging, it's now appreciated that effect sizes tend to be small, especially for differences between individuals, and therefore much larger samples are needed than the ones we were studying only a few years ago.[6]

It was reassuring, then, that other labs were finding convergent evidence for a role of the frontopolar cortex in metacognition. To have confidence in any finding in science, it is important to replicate it using multiple approaches. Coincidentally, Katsuki Nakamura's lab in Japan published a similar study in the same year as ours, but using measurements of brain function rather than structure. They found that the level of activity in the frontopolar cortex predicted differences in metacognition across individuals. A few years later, my collaborator Hakwan Lau replicated our results in his lab at Columbia University in New York, showing again that gray matter volume in the frontal pole was higher in individuals with better metacognition.[7]

The frontal pole, at the very front of the PFC, is thought to sit near the top of the kind of processing hierarchies that we encountered in the previous chapter. It is likely no coincidence that the frontopolar cortex is also one of the most expanded brain regions in humans compared to other primates. Anatomical studies of human and macaque monkey brains by researchers at Oxford University have found that many of the same subregions of the PFC can be reliably identified in both species. But they have also isolated unique differences in the lateral frontal pole, where the human brain appears to have acquired a new way station.[8]

Since we completed these initial studies, my lab has now amassed several data sets quantifying volunteers' metacognitive sensitivity on a variety of tasks. From this data, we are learning that there are surprisingly large and consistent differences in metacognition between individuals. One person may have limited insight into how well they are doing from one moment to the next, while another may have good awareness of whether they are likely to be right or wrong—even if they are both performing at the same level on the task. Another feature of metacognition is that, in controlled laboratory settings, it is relatively independent of other aspects of task performance. Your metacognition can still be in fine form as long as you recognize that you might be performing badly at the task (by having appropriately low confidence in your incorrect answers). This is the laboratory equivalent of recognizing a shoddy grasp of calculus or becoming painfully aware of disfluency in a new language. Self-awareness is often most useful for recognizing when we have done stupid things.

The general picture that emerges from research using these tools to quantify metacognition is that while people are often overconfident—thinking that they are better than others—they are also reasonably sensitive to fluctuations in their performance. Surveys routinely find that people think they are "better than average" in attributes ranging from driving skill to job performance to intelligence—an overconfident metacognitive bias. (Academics are some of the worst culprits: in one study, 94 percent of university professors in the United States rated their teaching performance as "above average"—a statistical impossibility!) But despite these generally inflated self-evaluations, we can still recognize when we have made a mistake on a test or put the car in the wrong gear.[9]

We have also found that metacognition is a relatively stable feature of an individual—it is trait-like. In other words, if you have good metacognition when I test you today, then you are also likely to have good metacognition when I test you tomorrow. The Argentinian neuroscientist Mariano Sigman refers to this as your metacognitive "fingerprint."[10] The trait-like nature of metacognition suggests that other features of people's personality, cognitive abilities, and mental health might play a role in shaping self-awareness. In one of our studies, Tricia Seow and Marion Rouault asked hundreds of individuals to fill in a series of questionnaires about their mood, anxiety, habits, and beliefs. From the pattern of their answers, we could extract a set of numbers for each individual that indexed where they fell on three core dimensions of mental health: their levels of anxiety and depression, their levels of compulsive behavior, and their levels of social withdrawal. Where people fell along these three dimensions predicted their metacognitive fingerprint, measured on a separate task. More anxious people tended to have lower confidence but heightened metacognitive sensitivity, whereas more compulsive people showed the reverse pattern. This result is consistent with the idea that how we think about ourselves may fluctuate in tandem with our mental health.[11]

As part of this experiment, we also included a short IQ test. We found that while IQ was consistently related to overall task performance, as expected, it was unrelated to metacognitive sensitivity. Because our sample contained almost one thousand individuals, it is likely that if a systematic relationship between IQ and metacognition had existed, then we would have detected it. Another piece of evidence for a difference between intelligence and self-awareness comes from a study I carried out in collaboration with neuropsychologist Karen Blackmon in New York, while I was a postdoctoral researcher at NYU. We found that patients who had recently had surgery to remove tumors from their anterior PFC had similar IQ to a control group but showed substantial impairments in metacognitive sensitivity. It is intriguing to consider that while both self-awareness and intelligence may depend on the PFC, the brain circuits that support flexible thinking may be distinct from those supporting thinking *about* thinking.[12]

Constructing Confidence

The abstract nature of metacognition makes sense when we consider that the uppermost levels of the prefrontal hierarchy have access to a broad array of inputs from elsewhere in the brain. They have a wide-angle lens, pooling many different sources of information and allowing us to build up an abstract model of our skills and abilities. This implies that brain circuits involved in creating human self-awareness transcend both perception and action—combining estimates of uncertainty from our sensory systems with information about the success of our actions. The two building blocks of self-awareness we encountered at the start of the book are becoming intertwined.

Emerging evidence from laboratory experiments is consistent with this idea. For instance, subtly perturbing brain circuits involved in action planning using transcranial magnetic stimulation can alter our confidence in a perceptual judgment, even though our ability to make such judgments remains unchanged. Similarly, the simple act of committing to a decision—making an action to say that the stimulus was A or B—is sufficient to improve metacognitive sensitivity, suggesting that our own actions provide an important input to computations supporting self-awareness.[13] The results of experiments with typists also show us that detecting our errors relies on keeping track of both the action we made (the key press) and the perceptual consequence of this action (the word that appears on the screen). If the experiment is rigged such that a benevolent bit of computer code kicks in to correct the typists' errors before they notice (just like the predictive text on a smartphone), they slow down their typing speed—suggesting that the error is logged somewhere in their brain—but do not admit to having made a mistake. In contrast, if extra errors are annoyingly inserted on the screen—for instance, the word "word" might be altered to "worz" on the fly—the typists gullibly accept blame for errors they never made.[14]

More broadly, we can only recognize our errors and regret our mistakes if we know what we should have done in the first place but didn't do. In line with this idea, the neuroscientists Lucie Charles and Stanislas Dehaene have found that the neural signatures of error monitoring disappear when the stimulus is flashed so briefly that it is difficult to see. This makes intuitive sense—if we don't see the stimulus, then we are just guessing, and there is no way for us to become aware of whether we have made an error. We can only consciously evaluate our performance when the inputs (perception) and outputs (action) are clear and unambiguous.[15]

The wide-angle lens supporting metacognition means that the current state of our bodies also exerts a powerful influence on how confident we feel about our performance. For instance, when images of disgusted faces—which lead to changes in pupil size and heart rate—are flashed very briefly on the computer screen (so briefly that they are effectively invisible), people's confidence in completing an unrelated task is subtly modulated. Similar cross talk between bodily states and self-awareness is seen when people are asked to name odors: they tend to be more confident about identifying smells they find emotionally laden (such as petrol or cloves) than those they rated as less evocative (such as vanilla or dill). We can think of these effects as metacognitive versions of the visual illusions we encountered in Chapter 1. Because different emotional and bodily states are often associated with feelings of uncertainty in daily life, manipulating emotions in the lab may lead to surprising effects on our metacognitive judgments.[16]

Together, these laboratory studies of metacognition tell us that self-awareness is subtly buffeted by a range of bodily and brain states. By pooling information arising from different sources, the brain creates a global picture of how confident it is in its model of the world. This global nature of metacognition endows human self-awareness with flexibility: over the course of a single day we may reflect on what we are seeing, remembering, or feeling, and evaluate how we are performing either at work or in a sports team. Evidence for this idea comes from studies that have found that people's metacognition across two distinct tasks—such as a memory task and a general-knowledge task—is correlated, even if performance on the two tasks is not. Being able to know whether we are right or wrong is an ability that transcends the particular problem we are trying to solve. And if metacognition relies on a general resource, this suggests that when I reflect on whether my memory is accurate, I am using similar neural machinery as when I reflect on whether I am seeing things clearly, even though the sources of information for these two judgments are very different. Data from my lab supports this idea. Using brain imaging data recorded from the PFC, we were able to train a machine learning algorithm to predict people's levels of confidence on a memory task (Which of two images do you remember?) using data on neural patterns recorded during a perception task (Which of two images is brighter?). The fact that we could do this suggests the brain tracks confidence using a relatively abstract, generic neural code.[17]

We can even think of personality features such as self-esteem and self-worth as sitting at the very top of a hierarchy of internal models, continuously informed by lower-level estimates of success in a range of endeavors, whether at work, at home, or on the sports field. In studies led by Marion Rouault, we have found that people's "local" confidence estimates in completing a simple decision-making task are used to inform their "global" estimates of how well they thought they were performing overall. These expectations also affect how lower-level self-monitoring algorithms respond to various types of error signal. In an impressive demonstration of this effect, Sara Bengtsson has shown that when people are primed to feel clever (by reading sentences that contain words such as "brightest"), brain responses to errors made in an unrelated task are increased, whereas priming people to feel stupid (by reading words such as "moron") leads to decreased error-related activity. This pattern of results is consistent with the primes affecting higher-level estimates about ourselves. When I feel like I should perform well (the "clever" context) but actually make errors, this is a bigger error in my prediction than when I expect to perform poorly.[18]

One appealing hypothesis is that the cortical midline structures we encountered in the previous chapter create a "core" or "minimal" representation of confidence in various mental states, which, via interactions with the lateral frontopolar cortex, supports explicit judgments about ourselves. There are various pieces of data that are consistent with the this idea. Using a method that combines EEG with fMRI, one recent study has shown that confidence signals are seen early on in the medial PFC, followed by activation in the lateral frontal pole at the time when an explicit self-evaluation is required. We have also found that the extent to which medial and lateral prefrontal regions interact (via an analysis that examines the level of correlation in their activity profiles) predicts people's metacognitive sensitivity.[19]

By carefully mapping out individuals' confidence and performance on various tasks, then, we are beginning to build up a detailed picture of our volunteers' self-awareness profile. We are also beginning to understand how self-awareness is supported by the brain. It seems that no two individuals are alike. But what is the origin of these differences? And is there anything we can do to change them?

Design Teams for Metacognition

The way our brains work is the result of a complex interaction between evolution (nature) and development (nurture). This is the case for even basic features of our mental and physical makeup; it's almost never the case that we can say something is purely "genetic." But it is useful to consider the relative strength of different influences. For many of the neuronal autopilots we encountered early in the book, genetic influences lead the way. Our visual systems can solve inverse problems at breakneck speed because they have been shaped by natural selection to be able to efficiently perceive and categorize objects in our environment. Over many generations, those systems that were better able to solve the problem of vision tended to succeed, and others died out.

But in other cases, the mechanisms sitting inside our heads are also shaped by our upbringing and education. For instance, our education systems create a cognitive process for reading—one that, if it goes to plan, results in neural systems configured for reading written words. Genetic evolution provides us with a visual system that can detect and recognize fine text, and our culture, parenting, and education does the rest.

Reading is an example of intentional design. We want to get our kids to read better, and so we send them to school and develop state educational programs to achieve this goal. But there are also examples of mental abilities that are shaped by society and culture in ways that are *not* intentional—that arise from the natural interactions between children and their parents or teachers. The Oxford psychologist Cecilia Heyes suggests that mindreading is one such process: we are (often unintentionally) taught to read minds by members of our social group, similar to the way that we are intentionally taught to read words. Mindreading is an example of a cognitive gadget that we acquire by being brought up in a community that talks about other people's mental states.[20]

Evidence for this idea comes from studies comparing mindreading across cultures. In Samoa, for instance, it is considered impolite to talk about mental states (how you are feeling or what you are thinking), and in these communities children pass mind-reading tests much later than children in the West, at around eight years of age. Perhaps the most persuasive evidence that mindreading is culturally acquired comes from an investigation of deaf users of Nicaraguan Sign Language. When this language was in its infancy, it only had rudimentary words for mental states, and people who learned it early on had a fairly poor understanding of false beliefs. But those who acquired it ten years later—when it was a mature language with lots of ways to talk about the mind—ended up being more adept at mindreading.[21]

A role for culture and parenting does not mean that genetics do not factor in, of course, or that there cannot be genetic reasons for failures of cultural learning. For instance, dyslexia has a heritable (genetic) component, which may lead to relatively general problems with integrating visual information. These genetic adjustments may go unnoticed in most situations, only becoming uncovered by an environment that places a premium on visual skills such as reading.

By comparing the similarity of identical twins (who share the same DNA) with the similarity of fraternal twins (who share only portions of DNA, in the same way as ordinary siblings), it is possible to tease out the extra contribution the shared genetic code makes to different aspects of the twins' mental lives. When this analysis was applied to mindreading (using data from more than one thousand pairs of five-year-old twins), the correlations in performance for identical and nonidentical twins were very similar. This result suggests that the main driver of variation in mindreading skill is not genetics, but the influence of a shared environment (sharing the same parents).[22]

Similarly, in the case of metacognition, a genetic starter kit may enable the implicit self-monitoring that is in place early in life, and then our parents and teachers take over to finish the job. Parents often help their children figure out what they are feeling or thinking by saying things like "Now you see me" and "Now you don't" in peekaboo, or suggesting that the child might be tired or hungry in response to various cries. As children grow up, they may continue to receive feedback that helps them understand what they are going through: sports coaches might explain to student athletes that feelings of nervousness and excitement mean that they are in the zone, rather than that they are unprepared.[23]

This extended period of development and learning perhaps explains why metacognition changes throughout childhood, continuing into adolescence. In a study I carried out in collaboration with Sarah-Jayne Blakemore's research group, we asked twenty-eight teenagers aged between eleven and seventeen to take the same metacognition test that Rimona and I developed: judging which image contained a brighter patch and rating their confidence in their judgment. For each volunteer, we then calculated how reliably their confidence ratings could tell apart their correct and incorrect judgments. We also measured how many correct judgments they made overall. We found that while age did not matter for the ability to choose the brighter patch, it did matter for their metacognition. Older teenagers were better at monitoring how well they were doing, reaching the highest level of self-awareness around the same time they were taking their A levels.[24]

The extended development of metacognition in adolescence is related to the fact that one of the key hubs for metacognition in the brain, the PFC, takes a long time to mature. A baby's brain has more connections between cells (known as synapses) than an adult's brain—around twice as many, in fact. Rather than building up the connections piece by piece, the brain gradually eliminates connections that are no longer required, like a sculpture emerging from a block of marble. Improvements in metacognition in childhood are strongest in individuals who show more pronounced structural changes in parts of the frontal lobe—in particular, the ventromedial PFC, the same region that we saw in the previous chapter is a nexus for both metacognition and mindreading. This protracted development of sensitivity to our own minds—through childhood and into our teenage years—is another piece of evidence that acquiring fully-fledged self-awareness is a long, arduous process, perhaps one that requires learning new tools for thinking.[25]

Losing Insight

Differences between individuals in the general population are relatively subtle, and large samples are needed in order to detect relationships between the brain and metacognition. But more extreme changes in self-awareness are unfortunately common in disorders of mental health. Neurologists and psychiatrists have different words to describe this phenomenon. Neurologists talk of anosognosia—literally, the absence of knowing. Psychiatrists refer to a patient's lack of insight. In psychiatry in particular it used to be thought that lack of insight was due to conscious denial or other kinds of strategy to avoid admitting to having a problem. But there is growing evidence that lack of insight may be due to the brain mechanisms responsible for metacognition being themselves affected by brain damage or disorders of mental health.[26]

A striking example of one such case was published in 2009 in the journal *Neuropsychologia*. It told the story of patient "LM," a highly intelligent sixty-seven-year-old lady, recently retired from a job in publishing. She had suddenly collapsed in a heap, unable to move her left leg and arm. Brain scans revealed that a stroke had affected the right side of her brain, leading to paralysis. When Dr. Katerina Fotopoulou, a clinician at UCL, went to examine her at hospital, she was able to move her right arm, gesture, and hold a normal conversation. But her left side was frozen in place.

Remarkably, despite her injury confining her to a hospital bed, LM believed she was just fine. Upon questioning, she batted away the challenge that she was paralyzed and breezily asserted that she could move her arm and clap her hands, attempting to demonstrate this by waving her healthy right hand in front of her body as if clapping an imaginary left hand. Despite being otherwise sane, LM had a view of herself markedly at odds with reality, and her family and doctors could not convince her otherwise. Her self-knowledge had been eroded by damage to her brain.[27]

In the case of neurological problems such as stroke or brain tumors, clinicians often describe the consequence of brain damage as attacking the very foundations of our sense of self. The British neurosurgeon Henry Marsh notes that "the person with frontal-lobe damage rarely has any insight into it—how can the 'I' know that it is changed? It has nothing to compare itself with."[28] This is sometimes referred to as the frontal lobe paradox. The paradox is that people with frontal lobe damage may have significant difficulties with everyday tasks such as cooking or organizing their finances, but because their metacognition is impaired, they are unable to recognize that they need a helping hand. Without metacognition, we may lose the capacity to understand what we have lost. The connection between our self-impression and the reality of our behavior—the reality seen by our family and friends—becomes weakened.[29]

Anosognosia is also common in various forms of dementia. Take the following fictional example: Mary is seventy-six years old and lives alone in a small town in Connecticut. Each morning, she walks to the nearby shops to pick up her groceries, a daily routine that until recently she carried off without a hitch. But then she begins to find herself arriving at the shop without knowing what she came for. She becomes frustrated that she didn't bother to write a list. Her daughter notices other lapses, such as forgetting the names of her grandchildren. While it is clear to her doctors that Mary has Alzheimer's disease, for her, like millions of others with this devastating condition, she is not aware of the havoc it is wreaking on her memory. In a cruel double hit, the disease has attacked not only brain regions supporting memory but also regions involved in metacognition. These impairments in metacognition have substantial consequences, such as a reluctance to seek help, take medication, or avoid situations (such as going out in the car) in which memory failure can be dangerous—all of which increase the risk of harm and the burden on families and caregivers. However, despite its significant clinical importance, metacognition is not typically part of a standard neuropsychological assessment. Aside from a handful of pioneering studies, a lack of metacognition is still considered only an anecdotal feature of dementia, acutely understood by clinicians but not routinely quantified or investigated as a consequence of the disease.[30]

It is not difficult to see how these gradual alterations in metacognition might ultimately lead to a complete loss of connection with reality. If I am no longer able to track whether my memory or perception is accurate, it may be hard to distinguish things that are real from things that I have just imagined. In fact, this kind of failure to discern reality from imagination is something we all experience. One of my earliest memories from my childhood is of being at the zoo with my grandparents, watching the elephants. This might be a true memory; but, equally, because I have heard over the years that my grandparents took me to the zoo, it could be an imagined memory that I now take as being real. Being able to tell the difference requires me to ask something about the nature of another cognitive process—in this case, whether my memory is likely to be accurate or inaccurate. The same kind of cognitive processes that I use to second-guess myself and question my memories or perceptions—the processes that underpin metacognition—are also likely to contribute to an ability to know what is real and what is not.[31]

Some of the most debilitating of these cases are encountered in schizophrenia. Schizophrenia is a common disorder (with a lifetime prevalence of around 1 percent) with an onset in early adulthood, and one that can lead to a profound disconnection from reality. Patients often suffer from hallucinations and delusions (such as believing that their thoughts are being controlled by an external source). Psychiatric disorders such as schizophrenia are also associated with changes in the brain. We can't necessarily pick up these changes on an MRI scan in the same way as we can a large stroke, but we can see more subtle and widespread changes in how the brain is wired up and the chemistry that allows different regions to communicate with each other. If this rewiring affects the long-range connections between the association cortex and other areas of the brain, a loss of self-awareness and contact with reality may ensue.

My former PhD adviser Chris Frith has developed an influential theory of schizophrenia that focuses on the deficit in self-awareness as being a root cause of many symptoms of psychosis. The core idea is that if we are unable to predict what we will do next, then we may reasonably infer that our actions and thoughts are being controlled by some external, alien force (sometimes literally). To the extent to which metacognition and mindreading share a common basis, this distortion in metacognitive modeling may also extend to others—making it more likely that patients with delusions come to believe that other people are intending to communicate with them or mean them harm.[32]

Psychologists have devised ingenious experiments for testing how we distinguish reality from imagination in the lab. In one experiment, people were given well-known phrases such as "Laurel and Hardy" and asked to read them aloud. Some of the phrases were incomplete—such as "Romeo and ?"—and in these cases they were asked to fill in the second word themselves, saying "Romeo and Juliet." In another condition, the subjects simply listened to another person reading the sentences. Later in the experiment, they were asked questions about whether they had seen or imagined the second word in the phrase. In general, people are good at making these judgments but not perfect—sometimes they thought they had perceived something when they had actually imagined it, and vice versa. When people answer questions about whether something was perceived or imagined, they activate the frontopolar cortex—the region of the brain that we have found to be consistently involved in metacognition. Variation in the structure of the PFC also predicts reality-monitoring ability, and similar neural markers are altered in people with schizophrenia.[33]

Understanding failures of self-awareness as being due to physical damage to the brain circuitry for metacognition provides us with a gentler, more humane perspective on cases in which patients begin to lose contact with reality. Rather than blaming the person for being unable to see what has changed in their lives, we can instead view the loss of metacognition as a consequence of the disease. Therapies and treatments targeted at metacognitive function may prove helpful in restoring or modulating self-awareness. Recent attempts at developing metacognitive therapies have focused on gradually sowing the seeds of doubt in overconfident or delusional convictions and encouraging self-reflection. Clinical trials of the effectiveness of one such approach have found that it showed a small but consistent effect in reducing delusions, and it has become a recommended treatment for schizophrenia in Germany and Australia.[34]

Cultivating Self-Awareness

We have seen that self-awareness is both grounded in brain function and affected by our social environment. While the initial building blocks of metacognition (the machinery for self-monitoring and error detection) may be present early in life, fully-fledged self-awareness continues to develop long into adolescence, and, just like the ability to understand the mental states of others, is shaped by our culture and upbringing. All these developmental changes are accompanied by extended changes to the structure and function of the prefrontal and parietal network that is critical for adult metacognition.

By the time we reach adulthood, most of us have acquired a reasonably secure capacity for knowing ourselves. We have also seen, though, that metacognition varies widely across individuals. Some people might perform well at a task but have limited insight into how well they are doing. Others might have an acute awareness of whether they are getting things right or wrong, even (or especially) when they are performing poorly. Because our self-awareness acts like an internal signal of whether we think we are making errors, it is easy to see how even subtle distortions in metacognition may lead to persistent under- or overconfidence and contribute to anxiety and stress about performance.

The good news is that metacognition is not set in stone. LM's case, for instance, had an unexpected and encouraging resolution. In one examination session, Dr. Fotopoulou used a video camera to record her conversations. At a later date, she asked LM to watch the tape. A remarkable change occurred:

As soon as the video stopped, LM immediately and spontaneously commented: "I have not been very realistic."

EXAMINER (AF): "What do you mean?"

LM: "I have not been realistic about my left side not being able to move at all."

AF: "What do you think now?"

LM: "I cannot move at all."

AF: "What made you change your mind?"

LM: "The video. I did not realize I looked like this."

This exchange took place in the space of only a few minutes, but LM's restored insight was still intact six months later. Seeing new information about her own body was sufficient to trigger a sudden transformation in self-awareness. This is, of course, only a single case report, and the dynamics of losing and gaining insight are likely to vary widely across different disorders and different people. But it contains an important lesson: metacognition is not fixed. Just as LM was able to recalibrate her view of herself, our own more subtle cases of metacognitive failure can also be shaped and improved. A good place to start is by making sure we are aware of the situations in which self-awareness might fail. Let's turn to these next.

5

AVOIDING SELF-AWARENESS FAILURE

There are three Things extreamly hard, Steel, a Diamond, and to know one's self.

—BENJAMIN FRANKLIN

We have almost reached the end of our tour through the neuroscience of self-awareness and the end of Part I of the book. We have seen that the fundamental features of brains—their sensitivity to uncertainty and their ability to self-monitor—provide a rich set of building blocks of self-awareness. We started with simple systems that take in information from the world and process it to solve the inverse problems inherent in perceiving and acting. A vast array of neural autopilots trigger continual adjustments to keep our actions on track. Much self-monitoring takes place unconsciously, without rising to the level of explicit self-awareness, and is likely to depend on a genetic starter kit that is present early in life. We also share many of these aspects of metacognition with other animals.

These building blocks form a starting point for understanding the emergence of fully-fledged self-awareness in humans. We saw that mutually reinforcing connections between social interaction, language, and an expansion of the capacity for recursive, deep, hierarchical models led the human brain to acquire a unique capacity for conscious metacognition—an awareness of our own minds. This form of metacognition develops slowly in childhood and, even as adults, is sensitive to changes in mental health, stress levels, and social and cultural environment.

In Part II of this book we are going to consider how this remarkable capacity for self-awareness supercharges the human mind. We will encounter its role in learning and education, in making complex decisions, in collaborating with others, and ultimately in taking responsibility for our actions. Before we dive in, though, I want to take a moment to extract three key lessons from the science of self-awareness.

Metacognition Can Be Misleading

First of all, it is important to distinguish between the *capacity* for metacognition and the *accuracy* of the self-awareness that results.

The capacity for metacognition becomes available to us after waking each morning. We can turn our thoughts inward and begin to think about ourselves. The accuracy of metacognition, on the other hand, refers to whether our reflective judgments tend to track our actual skills and abilities. We often have metacognitive accuracy in mind when we critique colleagues or friends for lacking self-awareness—as in, "Bill was completely unaware that he was dominating the meeting." Implicit in this critique is the idea that if we were to actually ask Bill to reflect on whether he dominated the meeting, he would conclude he did not, even if the data said otherwise.

As we have seen, there are many cases in which metacognition may lead us astray and become decoupled from objective reality. We saw that sometimes metacognition can be fooled by devious experimenters inserting or scrubbing away our errors before "we" have noticed. It is tempting to see these as failures of self-awareness. But this is just what we would expect from a system that is trying to create its best guess at how it is performing at any given moment from noisy internal data.

In fact, metacognition is likely to be even more susceptible to illusions and distortions than perception. Our senses usually remain in touch with reality because we receive constant feedback from the environment. If I misperceive my coffee cup as twice as large as it is, then I will likely knock it over when I reach out to pick it up, and such errors will serve to rapidly recalibrate my model of the world. Because my body remains tightly coupled to the environment, there is only so much slack my senses can tolerate. The machinery for self-awareness has a tougher job: it must perform mental alchemy, conjuring up a sense of whether we are right or wrong from a fairly loose and diffuse feedback loop. The consequences of illusions about ourselves are usually less obvious. If we are lacking in self-awareness, we might get quizzical looks at committee meetings, but we won't tend to knock over coffee cups or fall down flights of stairs. Self-awareness is therefore less moored to reality and more prone to illusions.

One powerful source of metacognitive illusions is known as fluency. Fluency is the psychologist's catch-all term for the ease with which information is processed. If you are reading these words in a well-lit, quiet room, then they will be processed more fluently than if you were struggling to concentrate in dim light. Such fluency colors how we interpret information. For instance, in a study of the stock market, companies with easy-to-pronounce names (such as Deerbond) were on average valued more highly than companies with disfluent names (such as Jojemnen). And because fluency can also affect metacognition, it may create situations in which we feel like we are performing well when actually we are doing badly. We feel more confident about our decisions when we act more quickly, even if faster responses are not associated with greater accuracy. Similarly, we are more confident about being able to remember words written in a larger font, even if font size does not influence our ability to remember. There are many other cases where these influences of fluency can lead to metacognitive illusions—metacognitive versions of the perceptual illusions we encountered earlier in the book.[1] As the Nobel prize–winning psychologist Daniel Kahneman points out:

> Subjective confidence in a judgment is not a reasoned evaluation of the probability that this judgment is correct. Confidence is a feeling, which reflects the coherence of the information and the cognitive ease of processing it. It is wise to take admissions of uncertainty seriously, but declarations of high confidence mainly tell you that an individual has constructed a coherent story in his mind, not necessarily that the story is true.[2]

In fact, the process of making metacognitive judgments can be thought of as the brain solving another kind of inverse problem, working out what to think about itself based on limited data. Just as our sensory systems pool information to mutually constrain and anchor our view of the world, a range of cues constrain and anchor our view of ourselves. Sometimes these inputs are helpful, but other times they can be hijacked. Just as illusions of perception reveal the workings of a powerful generative model that is trying to figure out what is "out there" in the world, metacognitive illusions created by false feedback, devious experimenters, and post-hoc justifications reveal the workings of the constructive process underpinning self-awareness. The fragility of metacognition is both a pitfall and an opportunity. The pitfall is that, as we have seen, self-awareness can be easily distorted or destroyed through brain damage and disorder. But the opportunity is that self-awareness can be molded and shaped by the way we educate our children, interact with others, and organize our lives.

Self-Awareness Is Less Common Than We Think

Another surprising feature of self-awareness is that it is often absent or offline. When a task is well learned—such as skilled driving or concert piano playing—the need to be aware of the details of what we are doing becomes less necessary. Instead, as we saw in our discussions of action monitoring, fine-scale unconscious adjustments are sufficient to ensure our actions stay on course. If lower-level processes are operating as they should—if all is well—then there is no need to propagate such information upward as a higher-level prediction error. Self-awareness operates on a need-to-know basis. This feature has the odd consequence that the absence of self-awareness is often more common than we think.

A useful analogy for the engagement of self-awareness is how problem-solving is typically handled in any big organization. If the problem can be tackled by someone relatively junior, then usually they are encouraged to take the initiative and resolve it without bothering their boss (and their boss's boss). In these cases, it's possible that the boss never becomes aware of the problem; the error signal was dealt with lower down the hierarchy and did not propagate up to the corner office. In these cases, there is a real sense in which the organization lacked metacognitive awareness of the problem (which may or may not be problematic, depending on whether it was handled effectively).

For another example, take the case of driving. It is quite common to travel several miles on a familiar route without much awareness of steering or changing gear. Instead, our thoughts might wander away to think about other worries, concerns, or plans. As the psychologist Jonathan Schooler notes, "We are often startled by the discovery that our minds have wandered away from the situation at hand." Self-awareness depends on neural machinery that may, for various reasons, become uncoupled from whatever task we are currently engaged in.[3]

Laboratory studies have shown that these metacognitive fade-outs are more common than we like to think—anywhere between 15 and 50 percent of the time, depending on the task. One workhorse experiment is known as the sustained attention to response task (SART). The SART is very simple, and very boring. A rapid series of numbers is presented on the screen, and people are asked to press a button every time they see a number except when they see the number 3, when they are supposed to inhibit their response. In parts of the task when people report their minds having wandered, their response times on the SART speed up, and they are more likely to (wrongly) hit buttons in response to the number 3.

This data suggests that when we mind-wander, we default to a mindless and repetitive stimulus-response mode of behaving. Perception and action are still churning away—they must be, in order to allow a response to the stimulus—but awareness of ourselves doing the SART has drifted off. This pattern is exacerbated under the influence of alcohol, when people become more likely to mind-wander and less likely to catch themselves doing so. Such mind-wandering episodes are also more likely when the task is well practiced, just as we would expect if self-awareness becomes redundant as skill increases.[4]

Mind-wandering, then, is a neat example of how awareness of an ongoing task might fade in and out. This does not mean that self-awareness is lost entirely; instead, it may become refocused on whatever it is we are daydreaming about. But this refocusing could mean that we are no longer aware of ourselves as actors in the world. A clever experiment by researchers at Weizmann Institute of Science in Israel attempted to measure what happens in the brain when awareness begins to fade in this way. They gave people two tasks to do. In one run of the fMRI scanner, subjects simply had to say whether a series of pictures contained animals or not. In another run, they saw the same series of images but were asked to introspect about whether the image elicited an emotional experience. The nice thing about this comparison is that the stimuli being presented (the images of animals) and the actions required (button presses) are identical in the two cases. The lower-level processes are similar; only the engagement of metacognition differs.

When the brain activity patterns elicited by these two types of scanner run were compared, a prefrontal network was more active in the introspection compared to the control condition, as we might expect if the PFC contributes to self-awareness. But the neatest aspect of the experiment was that the researchers also included a third condition, a harder version of the animal categorization task. The stimuli required much quicker responses, and the task was more stressful and engrossing. Strikingly, while this harder task *increased* the activation level in many brain areas (including parietal, premotor, and visual regions), it *decreased* activity in the prefrontal network associated with metacognition. The implication is that self-awareness had decreased as the task became more engaging. Similar neural changes may underpin the fading of self-awareness we experience when we are engrossed in a film or video game.[5]

Another factor that might lead to similar fade-outs in self-awareness, but for different reasons, is stress. Much is now known about the neurobiology of the stress response in both animals and humans, with one of the well-documented actions of stress hormones such as glucocorticoids being a weakening of the functions of the PFC. One implication is that metacognition may be one of the first functions to become compromised under stress. Consistent with this idea, in one study, people who showed greater cortisol release in response to a social stress test were also those who showed the most impaired metacognition. Similarly, in another experiment, simply giving people a small dose of hydrocortisone—which leads to a spike in cortisol over a period of a few hours—was sufficient to decrease metacognitive sensitivity compared to a group that received a placebo.[6]

A link between stress and lowered metacognition has some disturbing consequences. Arguably, we most need self-awareness at precisely the times when it is likely to be impaired. As the pressure comes on at work or when we are stressed by money or family worries, engaging in metacognition might reap the most benefits, enabling us to recognize errors, recruit outside help, or change strategy. Instead, if metacognition fades out under times of stress, it is more likely that we will ignore our errors, avoid asking for help, and plow on regardless.

Our second lesson from the science of self-awareness, then, is that the machinery for self-awareness might sometimes become disengaged from what we are currently doing, saying, or thinking. By definition, it is particularly hard to recognize when such fade-outs occur, because—as in the frontal lobe paradox—a loss of self-awareness also impacts the very functions we would need to realize its loss. The frequency of such self-awareness fade-outs is more common than we think.[7]

The Causal Power of Metacognition

A final lesson of the science of metacognition is that it has consequences for how we behave. Rather than self-awareness being a mere by-product of our minds at work, it has causal power in guiding our behavior. How do we know this?

One way of going about answering this question is to directly alter people's metacognition and observe what happens.

If how you feel about your performance changes, and these feelings play a causal role in guiding behavior, then your decision about what to do next—whether skipping a question or changing your mind—should also change. But if these feelings are epiphenomenal, with no causal role, then manipulating metacognition should have no effect.

The data so far supports a causal role for metacognition in guiding learning and decision-making. When people are given pairs of words to learn (such as "duck-carrot"), we can generate an illusion of confidence in having remembered the words by simply repeating the pairs during learning. Crucially, when people are made to feel more confident in this way, they are also less likely to choose the pairs again. The illusion of confidence is sufficient to make them believe further study is not necessary. Similar illusions of confidence about our decisions can be reliably induced using a trick known as a "positive evidence" manipulation. Imagine deciding which of two stimuli, image A or image B, has just been flashed on the computer screen. The images are presented in noise so it's difficult to tell A from B. If we now ramp up the brightness of both the image and the noise, the signal strength in favor of A or B increases, but the judgment remains objectively just as difficult (because the noise has also increased). Remarkably, this method is a reliable way of making people feel more confident, even if they are no more accurate in their choices. These heightened feelings of confidence affect how people behave, making them less likely to seek out new information or change their mind. The general lesson is that altering how we feel about our performance is sufficient to change our behavior, even if performance itself remains unchanged.[8]

Like any powerful tool, metacognition can be both creative and destructive. If our metacognition is accurate, it can supercharge how we go about our daily lives. Seemingly minor decisions made at a metacognitive level can have an outsize impact. For instance, I occasionally have to decide how much time to allocate to preparing a new lecture about my research. If I prepare all week, then I might increase my chances of giving a good talk, but this would be at the expense of actually doing the research that leads to invites to give talks in the first place! I need to have reasonable metacognition to know where to invest my time and energy, based on knowledge of my weaknesses in different areas. But if we act on metacognitive illusions, our performance can suffer. If I thought I was terrible at giving talks, I might spend all week investing unnecessary time practicing, leaving no time for other aspects of my job. Conversely, being overconfident and not preparing at all might result in an embarrassing failure.

Again, aircraft pilots give us a neat metaphor for the power of metacognitive illusions. Typically, when all is well, the different levels of self-monitoring of the aircraft are in alignment. At a lower level, the autopilot and instrument panels might inform the pilots that they are flying level at ten thousand feet, and they have no reason to disbelieve this. But in a particularly devastating situation known as the Coriolis illusion, in thick cloud pilots can sometimes think they are flying banked over when in fact their instruments (correctly) inform them that they are flying straight and level. Trying to correct the illusory bank can lead the aircraft to get into trouble. This factor was thought to play a role in the death of John F. Kennedy Jr. when his light aircraft crashed over Martha's Vineyard. Student pilots are now routinely instructed about the possibility of such illusions and told to always trust their instruments unless there is a very good reason not to.

We can protect ourselves from metacognitive illusions by taking a leaf out of the flying instructor's book. By learning about the situations in which self-awareness may become impaired, we can take steps to ensure we retain a reasonably clear-eyed view of ourselves. More specifically, we should be wary about placing too much trust in our metacognition if other sources (such as feedback from family and friends) tell us we are veering off course.

Like many things in science, when we begin to understand how something works, we can also begin to look for ways to harness it. In Part II, we are going to zoom in on the role that self-awareness plays in how we educate our children, make high-stakes decisions, work in teams, and augment the power of AI. This will not be a self-help guide with straightforward answers. It's rare that the science of self-awareness tells us we should do x but not y. But I hope that by understanding how self-awareness works—and, particularly, why it might fail—we can learn how to use it better.

PART II

✦

THE POWER OF REFLECTION

6

LEARNING TO LEARN

> Rather amazingly, we are animals who can think about any aspect of our own thinking and can thus devise cognitive strategies (which may be more or less indirect and baroque) aimed to modify, alter, or control aspects of our own psychology.
>
> —**ANDY CLARK,** *Supersizing the Mind*

From the Industrial Revolution onward, a dominant model in education was one of rote learning of facts and memorization—of capital cities, times tables, body parts, and chemical elements. In John Leslie's 1817 book *The Philosophy of Arithmetic*, he argued that children should be encouraged to memorize multiplication tables all the way up to 50 times 50. The schoolteacher Mr. Gradgrind in Charles Dickens's *Hard Times* agrees, telling us that "facts alone are wanted in life. Plant nothing else and root out everything else." The assumption was that the goal of education was to create people who can think faster and squirrel away more knowledge.

This approach may have achieved modest results in Victorian times. But in an increasingly complex and rapidly changing world, knowing *how* to think and learn is becoming just as important as *how much* we learn. With people living longer, having multiple jobs and careers, and taking up new hobbies, learning is becoming a lifelong pursuit, rather than something that stops when we leave formal schooling. As *The Economist* noted in 2017, "The curriculum needs to teach children how to study and think. A focus on 'metacognition' will make them better at picking up skills later in life."[1]

The consequences of being a good or poor learner ricochet throughout our lives. Cast your mind back to Jane, whom we encountered at the beginning of the book. To make smooth progress in her studying she needed to figure out what she knows and doesn't know and make decisions about what to learn next. These decisions may seem trivial, but they can be the difference between success and failure. If Jane's metacognition is good, then she will be able to effectively guide her learning. If, on the other hand, her metacognition is off, it won't matter if she is a brilliant engineer—she will be setting herself up for failure.

The lessons from Part I on how metacognition works take on critical importance in the classroom. Because metacognition sets the stage for how we learn, the payoff from avoiding metacognitive failure can be sizeable. In this chapter, we will explore how to apply what we know about how metacognition works to improve the way we make decisions about how, what, and when to study.

In the learning process, metacognition gets into the game in at least three places. We begin by forming beliefs about how best to learn and what material we think we need to focus on—what psychologists refer to as judgments of learning. Think back to when you were brushing up for a test in high school, perhaps a French exam. You might have set aside an evening to study the material and learn the various vocabulary pairs. Without realizing it, you were also probably making judgments about your own learning: How well do you know the material? Which word pairs might be trickier than others? Do you need to test yourself? Is it time to stop studying and head out with friends?

Once we have learned the material, we then need to use it—for an exam, in a dinner party conversation, or as a contestant on *Who Wants to Be a Millionaire?* Here, as we have seen in Part I, metacognition creates a fluctuating sense of confidence in our knowledge that may or may not be related to the objective reality of whether we actually know what we are talking about. Pernicious illusions of fluency can lead to dangerous situations in which we feel confident about inaccurate knowledge. Finally, after we've put an answer down on paper, additional metacognitive processing kicks into gear, allowing us to reflect on whether we might be wrong and perhaps to change our minds or alter our response. In what follows, we are going to look at each of these facets of self-awareness, and ask how we can avoid metacognitive failures.

The study of metacognition in the classroom has a long history. The pioneers of metacognition research were interested in how children's self-awareness affects how they learn. As a result, this chapter is able to draw on more "applied" research than other parts of the book—studies that have been carried out in classroom or university settings, and which have clear implications for how we educate our children. But it remains the case that the best route to improving our metacognition is by understanding how it works, and I aim to explain the recommendations of these studies through the lens of the model of self-awareness we developed in Part I.

Choosing How to Learn

As an undergraduate studying brain physiology, I was forced to learn long lists of cell types and anatomical labels. My strategy was to sit in the carrels of my college library with a textbook open, writing out the strange vocabulary on sheets of paper: Purkinje, spiny stellate, pyramidal; circle of Willis, cerebellar vermis, extrastriate cortex. I would then go through with a colored highlighter, picking out words I didn't know. In a final step, I would then (time and willpower permitting) transfer the highlighted items onto index cards, which I would carry around with me before the exam.

This worked for me, most of the time. But looking back, I had only blind faith in this particular approach to exam preparation. Deciding how to learn in the first place is a crucial choice. If we don't know how to make this decision, we might inadvertently end up holding ourselves back.

A common assumption is that each of us has a preferred learning style, such as being visual, auditory, or kinesthetic learners. But this is likely to be a myth. The educational neuroscientist Paul Howard-Jones points out that while more than 90 percent of teachers believe it is a good idea to tailor teaching to students' preferred learning styles, there is weak or no scientific evidence that people really do benefit from different styles. In fact, most controlled studies have shown no link between preferred learning style and performance. Yet the advice to tailor learning in this way has been propagated by august institutions such as the BBC and British Council.[2]

This widespread belief in learning styles may stem from a metacognitive illusion. In one study, fifty-two students were asked to complete a questionnaire measure of whether they prefer to learn through seeing pictures or written words. They were then given a memory test for common objects and animals, presented either as pictures or words. Critically, while the students were studying the word pairs, the researchers checked to see how confident the students felt by recording their judgments of learning. Their preferred learning style was unrelated to their actual performance: students who said they were pictorial learners were no better at learning from pictures, and those who said they were verbal learners were no better at learning from words. But it did affect their metacognition: pictorial learners felt more confident at learning from pictures, and verbal learners more confident about learning from words.[3]

The learning-styles myth, then, may be due to a metacognitive illusion: we feel more confident learning in our preferred style. The snag is that often the factors that make us learn better are also the things that make us *less* confident in our learning progress. Put simply, we might think we are better when using strategy A, whereas strategy B may actually lead to the best results.

Similar effects can be found when comparing digital and print reading. In one study, seventy undergraduates from the University of Haifa were asked to read a series of information leaflets about topics such as the advantages of different energy sources or the importance of warming up before exercise. Half of the time the texts were presented on a computer screen, and the other half they were printed out. After studying each text, students were asked how well they thought they would perform on a subsequent multiple-choice quiz. They were more confident about performing well after reading the information on-screen than on paper, despite performing similarly in both cases. This overconfidence had consequences: when they were allowed to study each passage for as long as they wanted, the heightened confidence for on-screen learning led to worse performance (63 percent correct versus 72 percent correct), because students gave up studying earlier.[4]

Another area in which metacognitive illusions can lead us astray is in deciding whether and how to practice—for instance, when studying material for an upcoming exam or test. A classic finding from cognitive psychology is that so-called spaced practice—reviewing the material once, taking a break for a day or two, and then returning to it a second time—is more effective for retaining information than massed practice, where the same amount of revision is crammed into a single session. Here again, though, metacognitive illusions may guide us in the wrong direction. Across multiple experiments, the psychologist Nate Kornell reported that 90 percent of college students had better performance after spaced, rather than massed, practice, but 72 percent of participants reported that massing, not spacing, was the more effective way of learning! This illusion may arise because cramming creates metacognitive fluency; it feels like it is working, even if it's not. Kornell likens this effect to going to the gym, only to choose weights that are too light and that don't have any training effect. It feels easy, but this isn't because you are doing well; it's because you have rigged the training session in your favor. In the same way that we want to feel like we've worked hard when we leave the gym, if we are approaching learning correctly, it usually feels like a mental workout rather than a gentle stroll.

Along similar lines, many students believe that just rereading their notes is the right way to study—just as I did when sitting in those library carrels, rereading my note cards about brain anatomy. This might feel useful and helpful, and it is probably better than not studying at all. But experiments have repeatedly shown that testing ourselves—forcing ourselves to practice exam questions, or writing out what we know—is more effective than passive rereading. It should no longer come as a surprise that our metacognitive beliefs about the best way to study are sometimes at odds with reality.[5]

Awareness of Ignorance

Once we have decided *how* to learn, we then need to make a series of microdecisions about *what* to learn. For instance, do I need to focus more on learning math or chemistry, or would my time be better spent practicing exam questions? This kind of metacognitive questioning does not stop being important when we leave school. A scientist might wonder whether they should spend more time learning new analysis tools or a new theory, and whether the benefits of doing so outweigh the time they could be spending on research. This kind of dilemma is now even more acute thanks to the rise of online courses providing high-quality material on topics ranging from data science to Descartes.

One influential theory of the role played by metacognition in choosing what to learn is known as the discrepancy reduction theory. It suggests that people begin studying new material by selecting a target level of learning and keep studying until their assessment of how much they know matches their target. One version of the theory is Janet Metcalfe's region of proximal learning (RPL) model. Metcalfe points out that people don't just strive to reduce the discrepancy between what they know and what they want to learn; they also prefer material that is not too difficult. The RPL model has a nice analogy with weight lifting. Just as the most gains are made in the gym by choosing weights that are a bit heavier than we're used to, but not so heavy that we can't lift them, students learn more quickly by selecting material of an intermediate difficulty.[6]

Both the discrepancy reduction and RPL models agree that metacognition plays a key role in learning by helping us monitor our progress toward a goal. Consistent with this idea, finely tuned metacognition has clear benefits in the classroom. For instance, one study asked children to "think aloud" while they were preparing for an upcoming history exam. Overall, 31 percent of the children's thoughts were classified as "metacognitive," as they were referring to whether they knew the material or not. Better readers and more academically successful students reported engaging in more metacognition.[7]

It stands to reason, then, that interventions to encourage metacognition might have widespread benefits for educational attainment. Patricia Chen and her colleagues at Stanford University set out to test this idea by dividing students into two groups prior to an upcoming exam. A control group received a reminder that their exam was coming up in a week's time and that they should start preparing for it. The experimental group received the same reminder, along with a strategic exercise prompting them to reflect on the format of the upcoming exam, which resources would best facilitate their studying (such as textbooks, lecture videos, and so on), and how they were planning to use each resource. Students in the experimental group outperformed those in the control group by around one-third of a letter grade: in a first experiment, the students performed an average of 3.6 percent better than controls, and in a second experiment, 4.2 percent better. Boosting metacognition also led to lower feelings of anxiety and stress about the upcoming exam.[8]

It may even be useful to cultivate what psychologists refer to as desirable difficulty, as a safeguard against illusions of confidence driven by fluency. For instance, scientists at RMIT University in Melbourne, Australia, have developed a new, freely downloadable computer font memorably entitled Sans Forgetica, which makes it harder than usual to read the words on the page. The idea is that the disfluency produced by the font prompts the students to think that they aren't learning the material very well, and as a result they concentrate and study for longer.[9]

Taken together, the current research indicates that metacognition is a crucial but underappreciated component of how we learn and study. What we think about our knowledge guides what we study next, which affects our knowledge, and so on in a virtuous (or sometimes vicious) cycle. This impact of metacognition is sometimes difficult to grasp and is not as easy to see or measure as someone's raw ability at math or science or music. But the impact of metacognition does not stop when learning is complete. As we saw in the case of Judith Keppel's game-show dilemma, it also plays a critical role in guiding how we *use* our newly acquired knowledge. This hidden role of metacognition in guiding our performance may be just as, if not more, important than raw intelligence for success on tests and exams. Let's turn to this next.

How We Know That We Know

Each year, millions of American high school juniors and seniors take the SAT, previously known as the Scholastic Aptitude Test. The stakes are high: the SAT is closely monitored by top colleges, and even after graduation, several businesses such as Goldman Sachs and McKinsey & Company want to know their prospective candidates' scores. At first glance, this seems reasonable. The test puts the students through their paces in reading, writing, and arithmetic and sifts them according to ability. Who wouldn't want the best and the brightest to enroll in their college or graduate program? But while raw ability certainly helps, it is not the only contributor to a good test score. In fact, until 2016, metacognition was every bit as important.

To understand why, we need to delve into the mechanics of how the SAT score was calculated. Most of the questions on the SAT are multiple-choice. Until 2016 there were five possible answers to each question, one of which is correct. If you were to randomly guess with your eyes closed throughout the test, you should expect to achieve a score of around 20 percent, not 0 percent. To estimate a true ability level, the SAT administrators therefore implemented a correction for guessing in their scoring system. For each correct answer, the student received one point. But for each incorrect answer, one-quarter of a point was deducted. This ensured that the average expected score if students guessed with their eyes closed was indeed 0.

However, this correction had an unintended consequence. The student could now strategically regulate his or her potential score rather than simply volunteer an answer for each question. If they had low confidence in a particular answer, they could skip the question, avoiding a potential penalty. We have already seen from the studies of animal metacognition in Part I that being able to opt out of a decision when we are not sure of the answer helps us achieve higher performance, and that this ability relies on effective estimation of uncertainty. Students with excellent metacognition would adeptly dodge questions they knew they would fail on and would lose only a few points. But students with poor metacognition—even those with above-average ability—might rashly ink in several incorrect responses, totting up a series of quarter-point penalties as they did so.[10]

The accuracy of metacognition may even make the difference between acing and failing a test. In one laboratory version of the SAT, volunteers were asked to answer a series of general-knowledge questions such as "What was the name of the first emperor of Rome?" If they weren't sure of the answer, they were asked to guess, and after each response the volunteers rated their confidence on a numerical scale. As expected, answers held with higher confidence were more likely to be correct. Now, armed with confidence scores for every question, the researchers repeated the test under SAT-style conditions: questions could be skipped, and there was a penalty for wrong answers. People tended to omit answers they originally held with lower confidence, and did so to a greater degree as the penalty for wrong responses went up, thus improving their score. But this same strategy had disastrous consequences when their metacognition was weak. In a second experiment, quiz questions were carefully chosen that often led to metacognitive illusions of confidence, such as "Who wrote the Unfinished Symphony?" or "What is the capital of Australia?" (the answers are Schubert, not Beethoven, and Canberra, not Sydney). For questions such as these, people's misplaced confidence leads them to volunteer lots of incorrect answers, and their performance plummets.[11]

Beneficial interactions between metacognition and test-taking performance may explain why metacognition plays a role in boosting educational success over time. One recent study looked at how metacognition and intelligence both developed in children aged between seven and twelve years old. Crucially, the same children came back to the lab for a subsequent assessment three years later (when the children were aged between nine and fifteen). From this rare longitudinal data, it was possible to ask whether the level of metacognition measured at age seven predicted a child's intelligence score at age nine and vice versa. While there were relatively weak links between metacognition and IQ at any one point in time (consistent with other findings of the independence of metacognition and intelligence), having good metacognition earlier in life predicted higher intelligence later on. An elegant explanation for this result is that having good metacognition helps children know what they don't know, in turn guiding their learning and overall educational progress. In line with this idea, when people are allowed to use metacognitive strategies to solve an IQ test, the extent to which they improve their performance by relying on metacognition is linked to real-life educational achievement.[12]

What is clear is that the scores from SAT-style tests are not only markers of ability, but also of how good individuals are at knowing their own minds. In using such tests, employers and colleges may be inadvertently selecting for metacognitive abilities as well as raw intelligence. This may not be a bad thing, and some organizations even do it on purpose: the British Civil Service graduate scheme asks potential candidates to rate their own performance during the course of their entrance examinations and takes these ratings into account when deciding whom to hire. The implication is that the Civil Service would prefer to recruit colleagues who are appropriately aware of their skills and limitations.

Believing in Ourselves

There is one final area in which metacognition plays a critical role in guiding our learning: creating beliefs about our skills and abilities. We have already seen in Part I that confidence is a construction and can sometimes be misleading—it does not always track what we are capable of. We might believe we will be unable to succeed in an upcoming exam, in a sports match, or in our career, even if we are more than good enough to do so. The danger is that these metacognitive distortions may become self-fulfilling prophecies. Put simply, if we are unwilling to compete, then we have no way of winning.

One of the pioneers in the study of these kinds of metacognitive illusions was the social psychologist Albert Bandura. In a series of influential books, Bandura outlined how what people believe about their skills and abilities is just as, if not more, important for their motivation and well-being than their objective abilities. He referred to this set of beliefs as "self-efficacy" (our overall confidence in our performance, closely related to metacognitive bias). He summarized his proposal as follows: "People's beliefs in their efficacy affect almost everything they do: how they think, motivate themselves, feel and behave." By subtly manipulating how people felt about their abilities on an upcoming task, laboratory experiments have borne out this hypothesis. Illusory boosts in self-efficacy indeed lead people to perform better, and persist for longer, at challenging tasks, whereas drops in self-efficacy lead to the opposite.[13]

One well-researched aspect of self-efficacy is children's beliefs about being able to solve math problems. In one set of longitudinal studies, children's beliefs about their abilities at age nine affected how they performed at age twelve, even when differences in objective ability were controlled for. The implication is that self-efficacy drove attainment, rather than the other way around. Because these beliefs can influence performance, any gender disparity in mathematics self-efficacy is a potential cause of differences between boys and girls in performance in STEM subjects. A recent worldwide survey showed that 35 percent of girls felt helpless when doing math problems, compared to 25 percent of boys. This difference was most prominent in Western countries such as New Zealand and Iceland, and less so in Eastern countries including Malaysia and Vietnam. It is not hard to imagine that systematic differences in self-efficacy may filter through into disparities in performance, leading to a self-reinforcing cycle of mathematics avoidance, even though girls start off no less able than boys.[14]

These effects of self-efficacy continue to echo into our adult lives. In social settings, such as at work and school, measures of women's confidence in their abilities are often lower than men's. (Interestingly, this difference disappears in our laboratory studies, in which we measure people's metacognition in isolation.) In their book *The Confidence Code*, Katty Kay and Claire Shipman describe a study conducted on Hewlett-Packard employees. They found that women applied for promotions when they believed they met 100 percent of the criteria, while men applied when they believed they met only 60 percent—that is, men were willing to act on a lower sense of confidence in their abilities. It is easy to see how this difference in confidence can lead to fewer promotions for women over the long run.[15]

On other occasions, though, lower self-efficacy can be adaptive. If we are aware of our weaknesses, we can benefit more from what psychologists refer to as offloading—using external tools to help us perform at maximum capacity. We often engage in offloading without thinking. Consider how you effortlessly reflect on whether you will be able to remember what you need to buy at the shop or if you need to write a list. Being self-aware of the limitations of your memory allows you to realize, "This is going to be tough." You know when you are no longer able to remember the items and need a helping hand.

Allowing people to offload typically improves their performance in laboratory tests, compared to control conditions where the offloading strategy is unavailable. This boost depends on estimates of self-efficacy. To know when to offload, people need to first recognize that their memory or problem-solving abilities might not be up to the job. People who have lower confidence in their memory are more likely to spontaneously set reminders, even after controlling for variation in objective ability. And this ability to use external props when things get difficult can be observed in children as young as four years old, consistent with self-efficacy and confidence guiding how we behave from an early age.[16]

To zero in on this relationship between metacognition and offloading, my student Xiao Hu and I set up a simple memory test in which people were asked to learn unrelated pairs of words, such as "rocket-garden" or "bucket-silver." The twist was that our participants were also given the opportunity to store any word pairs they weren't sure they could remember in a computer file—the lab version of writing a shopping list. When we then tested their memory for the word pairs a little later on, people naturally made use of the saved information to help them get the right answers, even though accessing the stored words carried a small financial cost. Critically, the volunteers in our study tended to use the stored information only when their confidence in their memory was low, demonstrating a direct link between fluctuations in metacognition and the decision to recruit outside help. Later in the book, we will see that this role of self-awareness in helping us know when and how to rely on artificial assistants is becoming more and more important, as our technological assistants morph from simple lists and notes to having minds of their own.[17]

Teaching Ourselves to Learn

So far, we have considered the role of metacognition in helping us know when we know and don't know and in guiding us toward the right things to learn next. But we have seen in Part I that metacognition (thinking about ourselves) and mindreading (thinking about others) are tightly intertwined. The human brain seems to use similar machinery in the two cases, just with different inputs. This suggests that simply thinking about what other people know (and what we think they should know) may feed back and sharpen our own understanding of what we know and don't know. As Seneca said, "While we teach, we learn."

The role of mindreading in teaching can already be seen in careful observations of young children. In one study, children aged between three and five were asked to teach a game to some puppets who were struggling with the rules. One puppet played the game perfectly, while others made mistakes. The older children tailored their teaching precisely, focusing on the needs of the puppets who made errors, whereas the younger children were more indiscriminate. This result is consistent with what we saw in Part I: around a critical age of about four years old, children become increasingly aware of other people's mental states. This awareness can facilitate teaching by allowing children to understand what others do and do not know.[18]

Children also intuitively know what needs to be taught and what doesn't. For instance, when preschool children were asked to think about someone who had grown up on a deserted island, they realized that the island resident could discover on their own that it's not possible to hold your breath for a whole day or that when a rock is thrown in the air it falls down. But they also appreciated that the same person would need to be taught that the Earth is round and that bodies need vitamins to stay healthy. This distinction between knowledge that can be directly acquired and that which would need to be taught was already present in the youngest children in the study (five years old) and became sharper with age.[19]

Teaching others encourages us to make our beliefs about what we know and don't know explicit and forces us to reconsider what we need to do to gain more secure knowledge of a topic. Prompts to consider the perspective of others indeed seem to have side benefits for self-awareness. When undergraduates were asked to study material for a quiz, they performed significantly better if they were also told they would have to teach the same material to another person. And asking students to engage in an eight-minute exercise advising younger students on their studies was sufficient to prompt higher grades during the rest of that school year, compared to a control group who did not provide advice. In their paper reporting this result, University of Pennsylvania psychologist Angela Duckworth and her colleagues suggest that the advice givers may have experienced heightened self-awareness of their own knowledge as a result of the intervention. This result is striking not least because meaningful educational interventions are notoriously difficult to find, and often those that are said to work may not actually do so when subjected to rigorous scientific testing. That a short advice-giving intervention could achieve meaningful effects on school grades is testament to the power of virtuous interactions between teaching, self-awareness, and performance.[20]

We can understand why teaching and advising others can be so useful for our own learning by considering how it helps us avoid metacognitive illusions. When we are forced to explain things to others, there is less opportunity to be swayed by internal signals of fluency that might create unwarranted feelings of confidence in our knowledge. For instance, the "illusion of explanatory depth" refers to the common experience of thinking we know how things work (from simple gadgets to government policies), but, when we are asked to explain them to others, we are unable to do so. Being forced to make our knowledge public ensures that misplaced overconfidence is exposed. For similar reasons, it is easier for us to recognize when someone else is talking nonsense than to recognize that same flaw in ourselves. Indeed, when people are asked to explain and justify their reasoning when judging difficult logic problems, they become more critical of their own arguments when they think they are someone else's rather than their own. Importantly, they also become more discerning—they are also more likely to correct their initially wrong answer to a problem when it is presented as someone else's answer.[21]

One implication of these findings is that a simple and powerful way to improve self-awareness is to take a third-person perspective on ourselves. An experiment conducted by the Israeli psychologists and metacognition experts Rakefet Ackerman and Asher Koriat is consistent with this idea. Students were asked to judge both their own learning and the learning progress of others, relayed via a video link. When judging themselves, they fell into the fluency trap; they believed that spending less time studying was a signal of confidence. But when judging others, this relationship was reversed; they (correctly) judged that spending longer on a topic would lead to better learning.[22]

External props and tools can also provide a new perspective on what we know. Rather than monitoring murky internal processes, many of which remain hidden from our awareness, getting words down on the page or speaking them out loud creates concrete targets for self-reflection. I have experienced this firsthand while writing this book. When I wasn't sure how to structure a particular section or chapter, I found the best strategy was to start by getting the key points down on paper, and only then was I able to see whether it made sense. We naturally extend our minds onto the page, and these extensions can themselves be targets of metacognition, to be mused about and reflected upon just like regular thoughts and feelings.[23]

Creating Self-Aware Students

Metacognition is central to how we learn new skills and educate our children. Even subtle distortions in the way students assess their skills, abilities, and knowledge can make the difference between success and failure. If we underestimate ourselves, we may be unwilling to put ourselves forward for an exam or a prize. If we overestimate ourselves, we may be in for a nasty shock on results day. If we cannot track how much we have learned, we will not know what to study next, and if we cannot detect when we might have made an error, we are unlikely to circle back and revise our answers in the high-pressure context of an exam. All of these features of metacognition are susceptible to illusions and distortions.

There is room for optimism, though. When students are encouraged to adopt a third-person perspective on their learning and teach others, they are less likely to fall prey to metacognitive distortions. By learning more about the cognitive science of learning (such as the costs and benefits of note-taking or different approaches to studying), rather than putting faith in sometimes misleading feelings of fluency, we can minimize these metacognitive errors.

Paying attention to the effects of metacognition is likely to have widespread benefits at all levels of our educational system, creating lean and hungry students who leave school having learned how to learn rather than being overstuffed with facts. There have been laudable efforts to improve metacognition in schools. But unfortunately these studies have not yet included the objective metrics of self-awareness we encountered in Part I, so we often do not know if they are working as intended. A good start will be to simply measure metacognition in our classrooms. Are we cultivating learners who know what they know and what they don't know? If we are not, we may wish to shift to an Athenian model in which cultivating self-awareness becomes just as prized as cultivating raw ability.

At higher levels of education, a broader implication is that lifelong teaching may in turn facilitate lifelong learning. The classical model in academia is that centers for teaching and learning should be collocated with centers for the discovery of new knowledge. Teaching and research are mutually beneficial. Unfortunately, this symbiosis is increasingly under threat. In the United States, the rise of adjunct teaching faculty without the status and research budgets of regular professors is decoupling teaching duties from research. In the UK, there is an increasing division between the research and teaching tracks, with junior academics on fellowships urged by their funders to protect their time for research. Too many teaching and administration responsibilities can indeed derail high-quality research—there is a balance to be struck. But I believe that all researchers should be encouraged to teach, if only to prompt us to reflect on what we know and don't know.

After we leave school, we may no longer need to hone our exam technique or figure out how best to study. But we are still faced with numerous scenarios where we are asked to question what we know and whether we might be wrong. In the next chapter, we are going to examine how self-awareness seeps into the choices and decisions we go on to make in our adult lives. We will see that the role of metacognition goes well beyond the classroom, affecting how we make decisions, work together with others, and take on positions of leadership and responsibility.

7

DECISIONS ABOUT DECISIONS

Even a popular pilot has to be able to land a plane.

—**KATTY KAY AND CLAIRE SHIPMAN,** *The Confidence Code*

In 2013, Mark Lynas underwent a radical change of mind about an issue that he had been passionate about for many years. As an environmental campaigner, he had originally believed that in creating genetically modified (GM) foods, science had crossed a line by conducting large-scale experiments on nature behind closed doors. He was militant in his opposition to GM, using machetes to hack up experimental crops and removing what he referred to as "genetic pollution" from farms and science labs around the country.

But then Lynas stood up at the Oxford Farming Conference and confessed that he had been wrong. The video of the meeting is on YouTube, and it makes for fascinating viewing. Reading from a script, he calmly explains that he had been ignorant about the science, and that he now realizes GM is a critical component of a sustainable farming system—a component that is saving lives in areas of the planet where non-GM crops would otherwise have died out due to disease. In an interview as part of the BBC Radio 4 series *Why I Changed My Mind*, Lynas comments that his admission felt "like changing sides in a war," and that he lost several close friends in the process.[1]

Lynas's story fascinates us precisely because such changes of heart on emotive issues are usually relatively rare. Instead, once we adopt a position, we often become stubbornly entrenched in our worldview and unwilling to incorporate alternative perspectives. It is not hard to see that when the facts change or new information comes to light this bias against changing our minds can be maladaptive or even downright dangerous.

But the lessons from Part I of this book give us reason for cautious optimism that these biases can be overcome. The kind of thought processes supported by the neural machinery for metacognition and mindreading can help jolt us out of an overly narrow perspective. Every decision we make, from pinpointing the source of a faint sound to choosing a new job, comes with a degree of confidence that we have made the right call. If this confidence is sufficiently low, it provides a cue to change our minds and reverse our decision. By allowing us to realize when we might be wrong, just as Lynas had an initial flicker of realization that he might have got the science backward, metacognition provides the mental foundations for subsequent changes of mind. And by allowing us to infer what other people know, and whether their knowledge is likely to be accurate or inaccurate, a capacity for mindreading ensures we benefit from the perspective and advice of others when working together in teams and groups. In this chapter, we are going to take a closer look at how metacognition enables (or disables) this ability to change our minds.[2]

To Change or Not to Change

We have already seen how Bayes's theorem provides us with a powerful framework for knowing when or whether to change our minds about a hypothesis. Consider the trick dice game from Chapter 1. If after multiple rolls of the dice we have seen only low numbers, we can be increasingly confident about a hypothesis that the trick die is showing a 0. In such a situation, one anomalous roll that returns a higher number such as a 10 should not affect our view, as this can still be consistent with the trick die showing a 0 (and the regular dice showing two 5s, or a 4 and a 6). More generally, the more confident a rational Bayesian becomes about a particular hypothesis, the less likely she should be to change her mind.

We have devised laboratory experiments to test this prediction and examine how the processing of new information is handled at a neural level. In these studies, we have extended our usual measure of metacognition—getting people to make difficult decisions about what they see on a computer screen—by allowing volunteers to see additional information after they have made a decision. In one version of this experiment, people are asked to say which way a cloud of noisy dots is moving. They are then shown another cloud of dots that is always moving in the same direction as the first, but which might be more or less noisy, and are then asked how confident they feel about their original decision.

A Bayesian observer solves this task by summing up the evidence samples (specifically, the sum of the ratio of the log-probability of each hypothesis) obtained before and after a decision, and then compares it to the decision that was actually made. By running computer simulations of these equations, we were able to make a specific prediction for the pattern of activity we should see in regions of the brain involved in revising our beliefs. If you were correct, the new evidence samples serve to confirm your initial choice, and the probability of being right goes up (we didn't try to trick our participants in this experiment). But if you were incorrect, then the new evidence samples disconfirm your initial choice, and the probability of being right goes down. The activity of a brain region engaged in updating our beliefs based on new evidence, then, should show opposite relationships with the strength of new evidence, depending on whether we were initially right or wrong.

By scanning people with fMRI while they are doing this task, we have been able to identify activity patterns in the brain that show precisely this neural signature. We were most interested in the cases in which people got their initial decision about the dots wrong, but then were presented with more information that they could use to reverse their choice. These results show that when the new information comes to light—when the new patch of dots comes on the screen—it triggers activation in the dorsal anterior cingulate cortex, and this activation pattern shows the signature predicted by the Bayesian model. The dACC is the same region that we saw in Chapter 2 that is involved in error detection. Our study of changes of mind suggests a more nuanced interpretation of the role of the dACC in metacognition—rather than simply announcing that we have made an error, it instead tracks how much we should update our beliefs based on new evidence.[3]

These results indicate that, just as tracking uncertainty is important for perceiving the world, it is also important for striking the right balance between being fixed and flexible in our beliefs. We can adapt our cake-batter analogy of Bayes's theorem to explain how uncertainty shapes whether or not we change our minds. Recall that the cake batter represents the data coming in, and the mold represents our prior belief about the world. Imagine that now the cake batter is of a medium consistency—not too runny and not too solid—and that we have two different molds, one made of thin, flexible rubber and another made of hard plastic. The flexibility of the mold represents how certain we are about our current beliefs about the world—for instance, whether the trick die is a 0 or a 3, or whether the dots are moving to the left or right. Thinner rubber—less confident beliefs—will tend to conform to the weight of the incoming batter, obliterating the original shape of the mold, whereas the hard plastic retains its shape, and you end up with a mold-shaped cake. The hard plastic is the equivalent of a confident belief: it retains its shape regardless of what the data throws at it.

This is all well and good, and a rational Bayesian can gracefully trade off current beliefs against new data. But the snag is that, as we have seen in Part I, our confidence may become unmoored from the accuracy of our models of the world. If we are overconfident, or think we have more reliable information than we do, we run the risk of not changing our minds when we should. In contrast, if we are underconfident, we may remain indecisive even when the way forward is clear. More generally, poor metacognition can leave us stuck with decisions, beliefs, and opinions that we should have reversed or discarded long ago.

The impact of confidence on changes of mind was the subject of a series of experiments led by Max Rollwage, a former PhD student in my group. He devised a variant of our task in which people were shown a cloud of moving dots on a computer screen and asked to decide whether they were moving to the left or right. After their initial choice, they had a chance to see the dots again—and, in some cases, this led people to change their minds. The clever part of the experiment was that by manipulating the properties of the first set of dots, Max could make people feel more confident in their initial choice, even if their performance remained unchanged (this is a version of the "positive evidence" effect we encountered in Part I). We found that these heightened feelings of confidence led to fewer changes of mind, exactly as we would expect if metacognitive feelings are playing a causal role in guiding future decisions about whether to take on board new information. However, now these feelings of confidence had become decoupled from the accuracy of people's decisions.[4]

Another prominent influence on how we process new evidence is confirmation bias. This refers to the fact that after making a choice, processing of new evidence that supports our choice tends to increase, whereas evidence that goes against our choice is downweighted. Confirmation bias has been documented in settings ranging from medical diagnoses to investment decisions and opinions on climate change. In one experiment, people were asked to bet on whether the list prices of various houses shown on a real estate website were more or less than a million dollars. They were then shown the opinion of a fictitious partner who either agreed or disagreed with their judgment about the house price and asked whether they would like to change their bet. The data showed that people became substantially more confident in their opinions when the partner agreed with them, but only slightly less confident when the partner disagreed. This asymmetry in the processing of new evidence was reflected in the activity profile of the dACC.[5]

At first glance, this pattern is not easy to square with Bayes's theorem, as a good Bayesian should be sensitive to new evidence regardless of whether they agree with it. But there is a further twist in this story. In our studies, Max found that this bias against disconfirmatory evidence is also modulated by how confident we feel in our initial decision. In this experiment, we used a technique known as magnetoencephalography (MEG), which can detect very small changes in the magnetic field around the heads of our volunteers. Because neurons communicate by firing tiny electrical impulses, it is possible to detect the telltale signs of this activity in subtle shifts in the magnetic field. By applying techniques borrowed from machine learning, it is even possible to decode features of people's thinking and decision-making from the spatial pattern of these changes in the magnetic field. In our experiment, we were able to decode whether people thought that a patch of moving dots was going to the left or to the right.

But we found that this decoding differed according to how confident people were in their decision. If they were highly confident, then any evidence that went against their previous decision was virtually un-decodable. It was as if the brain simply did not care about processing new evidence that contradicted a confident belief—a confidence-weighted confirmation bias.[6]

When we put all this data together, it leads to an intriguing hypothesis: perhaps seemingly maladaptive phenomena such as confirmation bias become beneficial when paired with good metacognition. The logic here is as follows: If holding high confidence tends to promote a bias toward confirmatory information, then this is OK, as long as I also tend to be correct when I am confident. If, on the other hand, I have poor metacognition— if I am sometimes highly confident and wrong—then on these occasions I will tend to ignore information that might refute my (incorrect) belief and have problems in building up a more accurate picture of the world.[7]

One consequence of this view is that there should be a tight coupling between people's metacognitive sensitivity and their ability to reconsider and reverse their decisions when they have made an error. We have directly tested this hypothesis in an experiment conducted online on hundreds of volunteers. People started off by completing a simple metacognition assessment: deciding which of two boxes on a computer screen contained more dots and rating their confidence in these decisions. We then gave them a variant of the information-processing task I described above. After making an initial decision, they saw another glimpse of the dots and were asked again to rate their confidence in their choice. In two separate experiments, we have found that people who have good metacognition in the first task tend to be the ones who are more willing to change their minds after making an error in the second task, demonstrating a direct link between self-awareness and more careful, considered decision-making.[8]

This power of metacognition in promoting a change in worldview can be naturally accommodated within a Bayesian framework. Our opinions can, as we have seen, be held with various levels of confidence or precision. For instance, I might have a very precise belief that the sun will rise tomorrow, but a less precise belief about the science on GM foods. The precision or confidence we attach to a particular model of the world is a metacognitive estimate, and as our cake-batter analogy highlights, how much we change our minds about something is crucially dependent on our current confidence. But the science of self-awareness also tells us that confidence is prone to illusions and biases. When metacognition is poor, being able to know when or whether to change our minds also begins to suffer.

From Perception to Value

Much of the research we have encountered so far in this book has focused on situations in which there is an objectively correct answer: whether a stimulus is tilted to the left or right, or whether a word was in a list we have just learned. But there is another class of decisions that are often closer to those we make in everyday life, and which don't have an objectively correct answer but are instead based on subjective preferences. Neuroscientists refer to these as value-based decisions, to contrast them against perceptual decisions. A perceptual decision would be whether I think the object on the kitchen table is more likely to be an apple or an orange, whereas a value-based decision would be whether I would prefer to eat that same apple or orange. The former can be correct or incorrect (I might have misperceived the apple as an orange from a distance, for instance) whereas it would be strange to say that I have incorrectly decided to eat the apple. No one can tell me I'm wrong about this decision; they instead should assume that I simply prefer apples over oranges. It would therefore seem strange to tell *myself* that I'm wrong about what I want. Or would it?

This was the question that I began to tackle with my friend and UCL colleague Benedetto De Martino in 2011, just after I moved to New York to start my postdoc. On trips back to London, and over multiple Skype calls, we debated and argued over whether it was even sensible for the brain to have metacognition about value-based choices. The key issue was the following: if people "knew" that they preferred the orange over the apple, surely they would have chosen it in the first place. There didn't seem to be much room for metacognition to get into the game.

The problem we studied was one familiar to behavioral economists. Say you are choosing dessert in a restaurant, and you have the opportunity to choose between two similarly priced ice creams, one with two scoops of vanilla and one of chocolate, and another with two scoops of chocolate and one of vanilla. If you choose the predominantly chocolate ice cream, we can infer that your internal ranking of preferences is for chocolate over vanilla; you have revealed your preference through your choice. If we were to then give you choices between all possible pairs and mixtures of ice creams, it would be possible to reconstruct a fairly detailed picture of your internal preferences for ice cream flavors just from observing the pattern of your choices.

We can make this situation more formal by assigning a numerical value to these internal (unobserved) preferences. Say I reveal through my choices that chocolate is worth twice as much to me as vanilla; then I can write that $U_{chocolate} = 2U_{vanilla}$, where U stands for an abstract quantity, the "utility" or value of the ice cream I will eventually eat (in these cases, we can think of utility as referring to all the subjective benefits I will accrue from the ice cream—including the taste, calories, and so on—minus the costs, such as worrying about a change in waistline). We can also define a sense of confidence, C, in our choice of flavor. It seems intuitive that when the difference in the utility between the two options becomes greater, then our confidence in making a good decision also increases. If I strongly prefer chocolate, I should be more confident about choosing one scoop of chocolate over one of vanilla, even more confident about choosing two scoops, and most confident of all about being given a voucher for unlimited purchases of chocolate ice cream. As the difference in value increases, the decision gets easier. We can write down this assumption mathematically as follows:

$$C \propto |U_A - U_B|$$

It says that our confidence is proportional to the absolute difference in value between the two options.

The problem is that this equation never allows us to think we have made an error—we can never be more confident about the option we didn't choose. This intuitive model, then, is inconsistent with people being able to apply metacognition to value-based choices; confidence is perfectly aligned with (and defined by) the values of the items we are deciding about. This seemed odd to us. The more we thought about it, the more we realized that metacognition about value-based choices was not only possible but central to how we live our lives. We began to see this kind of metacognition everywhere.

Consider choosing whether to take a new job. This is a value-based choice between option A (your existing job) and option B (the new job). After thinking carefully through the pros and cons—your colleagues, promotion opportunities, commute, and so on—you might come up with an overall sense that B is better. You decide on leaving, and carefully draft both resignation and acceptance letters. But then a wave of regret and second-guessing hits. Have you actually made the right choice? Wouldn't you have been better staying where you are? These are metacognitive thoughts about whether you have made a good decision—that is, metacognition about value-based choices. This kind of self-endorsement of our choices is a key aspect of decision-making, and it can have profound consequences for whether we decide to reverse or undo such decisions.

Together with our colleagues Neil Garrett and Ray Dolan, Benedetto and I set out to investigate people's self-awareness about their subjective choices in the lab. In order to apply the statistical models of metacognition that we encountered in Chapter 4, we needed to get people to make lots of choices, one after the other, and rate their confidence in choosing the best option—a proxy for whether they in fact wanted what they chose. We collected a set of British snacks, such as chocolate bars and crisps, and presented people with all possible pairs of items to choose between (hundreds of pairs in total). For instance, on some trials you might be asked to choose whether you prefer the Milky Way to the Kettle Chips, and on other trials whether you prefer the Lion Bar to the Twirl, or the Twirl to the Kettle Chips. We made sure that these choices mattered in several ways. First, one of the decisions was chosen at random to be played out for real, and people could eat the item they chose. Second, people were asked to fast for four hours before coming to the lab, so they were hungry. And third, they were asked to stay in the lab for an hour after the study, and the only thing they could eat was one of the snacks they chose in the experiment.

We next explored people's metacognition by applying statistical models that estimate the link between accuracy—whether they were actually right or wrong—and confidence. The snag in the value-based choice experiments was that it was difficult to define what "accurate" meant. How could we tell whether people really intended to choose the Lion Bar over the Twirl?

As a first attempt, we asked people to provide what amount of money they would be willing to pay for each snack after the experiment (again, to ensure people were incentivized to state their real willingness to pay, we made sure that this had consequences: they had more chances of getting a snack to eat if they indicated a higher price). For instance, you might bid £0.50 for the Lion Bar, but £1.50 for the Twirl, a clear statement that you prefer Twirls to Lion Bars. We then had both of the components we needed to estimate the equations above: people's confidence in each decision and a (subjective) value for each item.

The first thing we found in the data was that people tended to be more confident about easier decisions, just as the standard model predicts. What was more surprising, though, was that even when the value difference was the same—when two decisions were subjectively equally difficult—sometimes people's confidence was high, and sometimes it was low. When we dug into the data, we found that when people were highly confident, they were more likely to have chosen the snack they were willing to pay more for. But when they were less confident, they sometimes chose the snack that was less valuable to them. It seemed that people in our experiment were aware of making subjective errors—cases in which they chose the Twirl, but realized that, actually, they preferred the Lion Bar after all.

We also used fMRI to track the neural basis of people's decision-making process. We found that, in line with many other studies of subjective decisions, the values of different snacks were tracked by brain activity in the ventromedial PFC. This same region also showed higher activation when people were more confident in their choices. In contrast, the lateral frontopolar cortex—a brain region that, as we saw in Part I, is important for metacognitive sensitivity—tracked the confidence people had in their choices, but was relatively insensitive to their value. In other words, people have a sense of when they are acting in line with their own values, and this self-knowledge may share a similar neural basis to metacognition about other types of decisions.[9]

These experiments reveal that there is a real sense in which we can "want to want" something. We have an awareness of whether the choices we are making are aligned with our preferences, and this sense of confidence can be harnessed to ensure that, over the long run, we end up both wanting what we choose and choosing what we want. In fact, later on in our experiment, participants encountered the exact same pairs of snacks a second time, allowing us to identify cases in which they switched their choice from one snack to another. When people had low confidence in an initial choice, they were more likely to change their mind the second time around—allowing them to make choices that were, in the end, more in line with their preferences.[10]

A Delicate Balance

Our research on metacognition and changes of mind suggests that, in fact, being willing to admit to having low confidence in our decisions can often prove adaptive. By being open to change, we become receptive to new information that might contradict our existing view, just as in the case of Mark Lynas. And, as we have seen, this is most useful when our metacognition is accurate. We want to be open to changing our minds when we are likely to be wrong but remain steadfast when we are right. In this way, good metacognition nudges us into adopting a more reflective style of thinking, and protects us against developing inaccurate beliefs about the world.

For instance, consider the following question:

A bat and a ball cost £1.10 in total. The bat costs £1.00 more than the ball. How much does the ball cost?

The intuitive answer—one that is given by a high proportion of research participants—is 10p. But a moment's thought tells us that this can't be right: if the bat cost £1.00 more than 10p, it alone costs £1.10, which means that the bat and the ball together would cost £1.20. By working back through the sum, we find that, actually, the answer is 5p.

This question is part of the Cognitive Reflection Test (CRT) developed by the psychologist Shane Frederick. One difficulty with getting the right answers to questions such as this is that they have been designed to maximize metacognitive illusions. They generate a high feeling of confidence in knowing the answer despite accuracy being poor. And if our initial feeling of confidence is high, we might blurt out "10p" without pausing to reconsider or change our minds.[11]

People's scores on the CRT reliably predict performance in other endeavors that prize rational, reflective thought, including understanding of scientific topics, rejection of paranormal ideas, and the ability to detect fake news. One interpretation of these associations is that the CRT is tapping into a style of self-aware thinking that promotes the ability to know when we might be wrong and that new evidence is needed. These statistical associations remain even when controlling for general cognitive ability, suggesting that self-reflectiveness as measured by the CRT, like measures of metacognitive sensitivity, may be distinct from raw intelligence.[12]

Detailed studies of why CRT failures occur suggest that people would do better if they listened to initial feelings of low confidence in their answers and took time to reconsider their decisions. But there is another force at work that cuts in the opposite direction and tends to make us overconfident, irrespective of whether we are right or wrong. This is the fact that projecting confidence and decisiveness holds a subjective allure in the eyes of others. Despite all the benefits of knowing our limits and listening to feelings of low confidence, many of us still prefer to be fast, decisive, and confident in our daily lives—and prefer our leaders and politicians to be the same. What explains this paradox?

One clue comes from an elegant study by the political scientists Dominic Johnson and James Fowler. They set up a computer game in which large numbers of simulated characters were allowed to compete for limited resources. As in standard evolutionary simulations, those who won the competitions tended to acquire more fitness and be more likely to survive. Each of the characters also had an objective strength or ability that made them more or less likely to win the competition for resources. The twist was that, here, the decision about whether or not to compete for a resource was determined by the character's metacognitive belief about its ability—its confidence—rather than actual ability. And this confidence level could be varied in the computer simulations, allowing the researchers to create and study both underconfident and overconfident agents.

Intriguingly, in most scenarios, overconfident agents tended to do a bit better. This was especially the case when the benefit of gaining a resource was high, and when there was uncertainty about the relative strength of different agents. The idea is that overconfidence is adaptive because it encourages you to engage in fights in situations where you might have otherwise demurred. As the old saying goes, "You have to be in it to win it." In this way, the benefits of a dose of overconfidence for decision-making are similar to the benefits of self-efficacy for learning. It can become a self-fulfilling prophecy.[13]

People who are more confident indeed seem to achieve greater social status and influence. In one experiment in which people had to collaborate to pinpoint US cities on a map, overconfident individuals were perceived as more competent by their partners, and this overconfidence was associated with more respect and admiration. When videotapes of the experiment were rated afterward, more confident individuals spoke more, used a more assertive tone of voice, and exhibited a calm and relaxed demeanor. As the authors of this study wryly concluded, "Overconfident individuals more convincingly displayed competence cues than did individuals who were actually competent."[14]

Projecting decisiveness rather than caution is also liked and respected in our leaders and politicians, whereas admitting mistakes is often taken as a sign of weakness. From lingering too long over the menu at a restaurant to abrupt U-turns by politicians, flip-flopping does not have a good reputation. In the autumn of 2007, the incumbent prime minister of the UK, Gordon Brown, was enjoying high popularity ratings. He had just taken over the leadership of the Labour Party from Tony Blair, and he had deftly dealt with a series of national crises including terrorist plots. All indications were that he was going to coast to victory in an election—a poll that was his alone to call. But his very public decision to postpone going to the polls tarred him with a reputation for dithering and indecisiveness, and his authority began to crumble. In the 2004 US presidential election, the Democratic candidate, John Kerry, was similarly plagued by accusations of flip-flopping. In one famous remark, he tried to explain his voting record on funding for the military in the Middle East by saying, "I actually did vote for the $87 billion, before I voted against it."

There is a delicate balancing act here. The evolutionary simulations confirm a benefit of overconfidence in situations in which different individuals are competing for a limited resource. A subtle boost to confidence can also make us more competitive and likeable in the eyes of others. But these studies do not incorporate the potential downsides of overconfidence for monitoring our *own* decision-making. With too much overconfidence, as we have seen, we lose the benefits of an internal check on whether we are right or wrong.

So does this mean that we are damned if we do and damned if we don't? Do we have to choose between being confident but unreflective leaders or meek, introspective followers?

Luckily, there is a middle road, one that takes advantage of the benefits of being aware of our weaknesses while strategically cultivating confidence when needed. The argument goes as follows: If I have the capacity for self-awareness, then I can be ready and willing to acknowledge when I might be wrong. But I can also strategically bluff, upping my confidence when needed. Bluffing is only truly bluffing when we have some awareness of reality. If not, it is just blind overconfidence. (I remember playing poker as a student with a friend who was only just getting the hang of the rules. He went all in on a worthless hand and bluffed everyone into folding. This would have been a stunningly confident and impressive move, had he actually known that he was bluffing!)[15]

This kind of strategic metacognition requires a split in our mental architecture, a dissociation between the confidence that we feel privately and the confidence that we signal to others. For example, we may deliberately overstate our confidence in order to persuade others, or understate it in order to avoid responsibility for potentially costly errors. In a recent experiment conducted by Dan Bang, a postdoctoral researcher in my group, we have gotten a glimpse of the neural machinery that might coordinate strategic metacognition. Dan set up a scenario in which people needed to collaborate with another fictional "player" to jointly judge the direction of a cloud of randomly moving dots on the computer screen, similar to those we had used in our experiments on changes of mind. The twist was that the players were engineered to have different levels of confidence. Some players tended to be underconfident, and others tended to be overconfident. But the rules of the game said that the most confident judgment would be taken as the group's decision—similar to the person who shouts loudest in meetings, dominating the proceedings. This meant that when collaborating with less-confident players, it was useful to strategically reduce your confidence (to avoid dominating the decision), whereas when playing with more-confident ones, it was better to shout a bit louder to ensure your voice was heard.

The volunteers in our experiment got this immediately, naturally shifting their confidence to roughly match that of their partners. We then looked at the patterns of brain activity in the prefrontal regions we knew were important for meta-cognition. In the ventromedial PFC, we found activation that tracked people's private sense of confidence in their decision. It was affected by how difficult the judgment was, but not by who they were playing with. In contrast, in the lateral fronto-polar cortex, we found neural signals that tracked how much people needed to strategically adjust their confidence when playing with different partners. These findings also help us fur-ther understand the distinction between implicit and explicit metacognition we encountered in Part I. Implicit signals of confidence and uncertainty seem to be tracked at multiple stages of neural processing. But the ability to strategically use and communicate confidence to others may depend on brain networks centered on the frontopolar cortex—networks that are uniquely expanded in the human brain and take a while to mature in childhood.[16]

It takes courage to adopt a metacognitive stance. As we have seen, publicly second-guessing ourselves puts us in a vulnerable position. It is perhaps no surprise, then, that some of the world's most successful leaders put a premium on strategic metacogni-tion and prolonged, reflective decision-making. Effective leaders are those who both are aware of their weaknesses and can stra-tegically project confidence when it is needed. As Ray Dalio recounts in his best-selling book *Principles*, "This episode taught me the importance of always fearing being wrong, no matter how confident I am that I'm right."[17]

In 2017, Amazon's letter to its shareholders was the usual mix of ambitious goals and recounted milestones that characterizes many global companies. But it, too, stood out for its unusual focus on self-awareness: "You can consider yourself a person of high standards in general and still have debilitating blind spots. There can be whole arenas of endeavor where you may not even know that your standards are low or nonexistent, and certainly not world class. It's critical to be open to that likelihood." Amazon CEO Jeff Bezos practices what he preaches, being famous for his unusual executive meetings. Rather than engaging in small talk around the boardroom table, he instead mandates that executives engage in a silent thirty-minute "study session," reading a memo that one of them has prepared in advance. The idea is to force each person to think about the material, form their own opinion, and reflect on what it means for them and the company. With his shareholder letter and the unusual meeting setup, it is clear that Bezos places high value on individual self-awareness—not only being well-informed, but also knowing what you know and what you do not.[18]

For Bezos, self-awareness is important for the same reason that it is important to sports coaches. By gaining self-awareness, we can recognize where there is room for improvement. Further on in the 2017 memo, he notes, "The football coach doesn't need to be able to throw, and a film director doesn't need to be able to act. But they both do need to recognize high standards."

This focus on metacognition espoused by many successful individuals and companies is likely to be no accident. It enables them to be agile and adaptive, realizing their mistakes before it's too late and recognizing when they may need to improve. As we saw in the story of Charmides, the Greeks considered *sophrosyne*—the living of a balanced, measured life—as being grounded in effective self-knowledge. By knowing what we want (as well as knowing what we know), we can reflectively endorse our good decisions and take steps to reverse or change the bad ones. It's often useful to be confident and decisive and project an air of reassurance toward others. But when the potential for errors arises, we want leaders with good metacognition, those who are willing to quickly recognize the danger they are in and change course accordingly. Conversely, self-awareness helps us be good citizens in a world that is increasingly suffused with information and polarized in opinion. It helps us realize when we don't know what we want, or when we might need more information in order to figure this out.

Just as in the cases of Jane, Judith, and James at the start of the book, if I have poor metacognition about my knowledge or skills or abilities, my future job prospects, financial situation, or physical health might suffer. In these isolated cases, failures of self-awareness are unlikely to affect others. But, as we have seen, the impact of metacognition is rarely limited to any one individual. If I am working with others, a lack of self-awareness may result in network effects—effects that are difficult to anticipate at the level of the individual but can have a detrimental impact on teams, groups, or even institutions. In the next chapter, we are going to expand our focus from individuals making decisions alone to people working together in groups. We will see that effective metacognition not only is central to how we reflect on and control our *own* thinking, but it also allows us to broadcast our mental states to others, becoming a catalyst for human collaboration of all kinds.

8

COLLABORATING AND SHARING

> Consciousness is actually nothing but a network of connections between man and man—only as such did it have to develop: a reclusive or predatory man would not have needed it.
>
> —FRIEDRICH NIETZSCHE, *The Joyous Science*

The increasing specialization of human knowledge means that an ability to work together on a global scale has never been more important. It's rare for one individual to have all the expertise needed to build a plane, treat a patient, or run a business. Instead, humans have succeeded in all these endeavors largely thanks to an ability to collaborate, sharing information and expertise as and when required. This ability to share and coordinate with others requires us to keep track of who knows what. For instance, when my wife and I go on vacation, I know that she will know where to look for the sunscreen. Conversely, she knows that I will know where to find the beach towels.

We have already seen that many animals have an ability for implicit metacognition. But it seems likely that only humans have an ability for explicitly representing the contents of the minds of ourselves and others. This capacity for self-awareness mutually reinforces the usefulness of language. Our linguistic abilities are clearly central to our ability to collaborate and share ideas. But language is not enough. Human language would be no more than an elaborate version of primitive alarm calls were it not for our ability to broadcast and share our thoughts and feelings. Monkeys use a set of calls and gestures to share information about what is going on in the world, such as the location of food sources and predators. But they do not, as far as we know, use their primitive language to communicate mental states, such as the fact that they are feeling fearful or anxious about a predator. No matter how baroque or complex linguistic competence becomes, without self-awareness we cannot use it to tell each other what we are thinking or feeling.[1]

The central role of self-awareness in working with others has a downside, though. It means that effective collaboration often depends on effective metacognition, and, as we have seen, there are many reasons why metacognition may fail. Subtle shifts in self-awareness then become significant when magnified at the level of groups and societies. In this chapter, we will see how metacognition plays a pivotal role in our capacity to coordinate and collaborate with others in fields as diverse as sports, law, and science—and see how the science of self-awareness can supercharge the next wave of human ingenuity.

Two Heads Are Better than One

Consider a pair of hunters stalking a deer. They crouch next to each other in the long grass, watching for any sign of movement. One whispers to the other, "I think I saw a movement over to the left." The other replies, "I didn't see that—but I definitely saw something over there. Let's press on." Like many of us in this situation, we might defer to the person who is more confident in what they saw.

Confidence is a useful currency for communicating strength of belief in these scenarios. On sports fields around the world, professional referees are used to pooling their confidence to come to a joint decision. As a boy growing up in the north of England, like all my friends I was obsessed with football (soccer). I vividly remember the European Championship of 1996, which was notable for the "Three Lions" song masterminded by David Baddiel, Frank Skinner, and the Lightning Seeds. This record—which, unusually for a football song, was actually pretty good—provided the soundtrack to a glorious summer. It contained the memorable line, "Jules Rimet still gleaming / Thirty years of hurt." Jules Rimet refers to the World Cup trophy, and "thirty years" reminded us that England hadn't won the World Cup since 1966.

Those lyrics prompted me to find out what happened back in 1966 at the famous final against West Germany at Wembley in London. Going into extra time, the game was tied at 2–2, and the whole country was holding its breath. A home World Cup was poised for a fairytale ending. On YouTube, you can find old TV footage showing Alan Ball crossing the ball to Geoff Hurst, the England striker. Shooting from close range, Hurst hits the underside of the crossbar and the ball bounces down onto the goal line. The England players think Hurst has scored, and they wheel away in celebration.

It was the linesman's job to decide whether the ball has crossed the line for a goal. The linesman that day was Tofiq Bahramov, from Azerbaijan in the former Soviet Union. Bahramov's origins provide some extra spice to the story, because West Germany had just knocked the USSR out of the competition in the semifinals. Now Bahramov was about to decide the fate of the game while four hundred million television viewers around the world looked on. Without recourse to technology, he decided the ball was over the line, and England was awarded the crucial goal. Hurst later added another to make it 4–2 in the dying seconds, but by then fans were already flooding the field in celebration.

Today's professional referees are linked up by radio microphones, and those in some sports such as rugby and cricket have the ability to request an immediate review of the TV footage before making their decision. What would have been the outcome had Bahramov communicated a degree of uncertainty in his impression of what happened? It is not hard to imagine that, absent any other influence on the referee's decision process, a more doubtful linesman may have swung the day in Germany's favor.

As we have seen, however, sharing our confidence is only useful if it's a reliable marker of the accuracy of others' judgments. We would not want to work with someone who is confident when they are wrong, or less confident when they are right. This importance of metacognition when making collective decisions has been elegantly demonstrated in the lab in studies by Bahador Bahrami and his colleagues. Pairs of individuals were asked to view briefly flashed stimuli, each on their own computer monitor. Their task was to decide whether the first or second flash contained a slightly brighter target. If they disagreed, they were prompted to discuss their reasons why and arrive at a joint decision. These cases are laboratory versions of the situation in a crowded Wembley: How do a referee and linesman arrive at a joint assessment of whether the ball crossed the line?

The results were both clear-cut and striking. First, the researchers quantified each individual's sensitivity when detecting targets alone. This provides a baseline against which to assess any change in performance when decisions were made with someone else. Then they examined the joint decisions in cases where participants initially disagreed. Remarkably, in most cases, decisions made jointly were more accurate than equivalent decisions made by the best subject working alone. This is known as the "two heads are better than one" (2HBT1) effect and can be explained as follows: Each individual provides some information about where the target was. A mathematical algorithm for combining these pieces of information weighs them by their reliability, leading to a joint accuracy that is greater than the sum of its parts. Because people intuitively communicate their confidence in their decisions, the 2HBT1 effect is obtained.[2]

Pairs of decision makers in these experiments tend to converge on a common, fine-grain currency for sharing their confidence (using phrases such as "I was sure" or "I was very sure"), and those who show greater convergence also show more collective benefit. There are also many implicit cues to what other people are thinking and feeling; we don't always need to rely on what they say. People tend to move more quickly and decisively when they are confident, and in several languages confident statements are generally marked by greater intonation and louder or faster pronunciation. There are even telltale hints of confidence in email and social media. In one experiment simulating online messaging, participants who were more confident in their beliefs tended to message first, and this was predictive of how persuasive the message was. These delicate, reciprocal, and intuitive metacognitive interactions allow groups of individuals to subtly modulate each other's impression of the world.[3]

Mistaken Identification

It is perhaps rare that we are asked, explicitly, to state how confident we are in our beliefs. But there is one high-stakes arena in which the accuracy of our metacognitive statements takes on critical importance. In courts of law around the world, witnesses take the stand to declare that they saw, or did not see, a particular crime occur. Often, confident eyewitness reports can be enough to sway a jury. But given everything we have learned about metacognition so far, we should not be surprised if what people say does not always match up with the truth.

In February 1987, eighteen-year-old Donte Booker was arrested for an incident involving a toy gun. Recalling that a similar toy gun was involved in a still-unsolved rape case, a police officer put Booker's photograph into a photo lineup. The victim confidently identified him as the attacker, which, together with the circumstantial evidence of the toy gun, was enough to send Booker to jail for twenty-five years. He was granted parole in 2002 after serving fifteen years of his sentence, and he started his life over again while battling the stigma of being a convicted sex offender. It wasn't until January 2005 that DNA testing definitively showed that Booker was not the rapist. The real attacker was an ex-convict whose DNA profile was obtained during a robbery he committed while Booker was already in prison.

Cases such as these are unfortunately common. The Innocence Project, a US public-policy organization dedicated to exonerating people who were wrongfully convicted, estimates that mistaken identification contributed to approximately 70 percent of the more than 375 wrongful convictions in the United States overturned by post-conviction DNA evidence (note that these are only the ones that have been proven wrongful; the real frequency of mistaken identification is likely much higher). And mistaken identifications are often due to failures of metacognition: eyewitnesses often believe that their memory of an event is accurate and report unreasonably high confidence in their identifications. Jurors take such eyewitness confidence to heart. In one mock-jury study, different factors associated with the crime were manipulated, such as whether the defendant was disguised or how confident the eyewitness was in their recall of the relevant details. Strikingly, the confidence of the eyewitness was the most powerful predictor of a guilty verdict. In similar studies, eyewitness confidence has been found to influence the jury more than the consistency of the testimony or even expert opinion.[4]

All this goes to show that good metacognition is central to the legal process. If the eyewitness has good metacognition, then they will be able to appropriately separate out occasions when they might be mistaken (communicating lower confidence to the jury) from occasions when they are more likely to be right. It is concerning, therefore, that studies of eyewitness memory in the laboratory have shown that metacognition is surprisingly poor.

In a series of experiments conducted in the 1990s, Thomas Busey, Elizabeth Loftus, and colleagues set out to study eyewitness confidence in the lab. Busey and Loftus asked participants to remember a list of faces, and afterward to respond as to whether a face was previously on the list or was novel. Participants were also asked to indicate their confidence in their decision. So far, this is a standard memory experiment, but the researchers introduced an intriguing twist. Half of the photographs were presented in a dim light on the computer screen, whereas the other half were presented brightly lit. This is analogous to a typical police lineup—often a witness captures only a dim glimpse of an attacker at the crime scene but is then asked to pick him or her out of a brightly lit lineup. The results were clear-cut and unsettling. Increasing the brightness of the face during the "lineup" phase decreased accuracy in identification but increased subjects' confidence that they had got the answer right. The authors concluded that "subjects apparently believe (slightly) that a brighter test stimulus will help them, when in fact it causes a substantial decrease in accuracy."[5]

This influence of light levels on people's identification confidence is another example of a metacognitive illusion. People feel that they are more likely to remember a face when it is brightly lit, even if this belief is inaccurate. Researchers have found that eyewitnesses also incorporate information from the police and others into their recollection of what happened, skewing their confidence further as time passes.

What, then, can we do about these potentially systemic failures of metacognition in our courtrooms? One route to tackling metacognitive failure is by making judges and juries more aware of the fragility of self-awareness. This is the route taken by the state of New Jersey, which in 2012 introduced jury instructions explaining that "although some research has found that highly confident witnesses are more likely to make accurate identifications, eyewitness confidence is generally an unreliable indicator of accuracy." A different strategy is to identify the conditions in which people's metacognition is intact and focus on them. Here, by applying the science of self-awareness, there is some reason for optimism. The key insight is that witness confidence is typically a good predictor of accuracy at the time of an initial lineup or ID parade but becomes poorer later on during the trial, after enough time has elapsed and people become convinced about their own mistaken ID.[6]

Another approach is to provide information about an individual's metacognitive fingerprint. For instance, mock jurors find witnesses who are confident about false information less credible. We tend to trust people who have better metacognition, all else being equal. We have learned that when they tell us something with confidence, it is likely to be true. The problem arises in situations where it is not possible to learn about people's metacognition—when we are interacting with them only a handful of times, or if they are surrounded by the anonymity of the Internet and social media. In those scenarios, confidence is king.

By understanding the factors that affect eyewitnesses' self-awareness, we can begin to design our institutions to ensure their testimony is a help, rather than a hindrance, to justice. For instance, we could ask the witness to take a metacognition test and report their results to the judge and jury. Or we could ensure that numerical confidence estimates are routinely recorded at the time of the lineup and read back to the jury to ensure that they are reminded about how the witness felt about their initial ID, when metacognition tends to be more accurate. Putting strict rules and regulations in place is easier to do in situations such as courts of law. But what about in the messy back-and-forth of our daily lives? Is there anything we can do to ensure we do not fall foul of self-awareness failures in our interactions with each other?[7]

The Right Kind of Ignorance

If someone has poor metacognition, then their confidence in their judgments and opinions will often be decoupled from reality. Over time, we might learn to discount their opinions on critical topics. But if we are interacting with someone for the first time, we are likely to give them the benefit of the doubt, listening to their advice and opinions, and assuming that their metacognition is, if not superlative, at least intact. I hope that, by now, you will be more cautious before making this assumption—we have already seen that metacognition is affected by variation in people's stress and anxiety levels, for instance. Metacognitive illusions and individual variability in self-awareness mean that it is wise to check ourselves before taking someone's certainty as an indicator of the correctness of their opinions.

This is all the more important when we are working with people for the first (or perhaps only) time. When we have only one or two data points—for instance, a single statement of high confidence in a judgment—it is impossible to assess that person's metacognitive sensitivity. We cannot know whether that person has good metacognition and is communicating they are confident because they are actually likely to be correct, or whether they have poor metacognition and are asserting high confidence regardless.

The lawyer Robert Rothkopf and I have studied these kinds of mismatches in the context of giving legal advice. Rob runs a litigation fund that invests in cases such as class actions. His team often needs to decide whether or not to invest capital in a new case, depending on whether litigation is likely to succeed. Rob noted that often lawyers presenting cases for investment used verbal descriptions of the chances of success: phrases like "reasonable prospects" or a "significant likelihood" of winning. But what do they really mean by this?

To find out, Rob and I issued a survey to 250 lawyers and corporate clients around the world, asking them to assign a percentage to verbal descriptions of confidence such as "near certainty," "reasonably arguable," and "fair chance." The most striking result was the substantial variability in the way people interpreted these phrases. For instance, the phrase "significant likelihood" was associated with probabilities ranging from below 25 percent to near 100 percent.[8]

The implications of these findings are clear. First, verbal labels of confidence are relatively imprecise and cover a range of probabilities, depending on who is wielding the pen. Second, different sectors of a profession might well be talking past each other if they do not have a shared language to discuss their confidence in their beliefs. And third, without knowing something about the accuracy of someone's confidence judgments over time, isolated statements need to be taken with considerable caution. To avoid these pitfalls, Rob's team now requires lawyers advising them to use numerical estimates of probabilities instead of vague phrases. They have also developed a procedure that requires each member of the team to score their own predictions about a case independently before beginning group discussions, to precisely capture individual confidence estimates.[9]

Scientists are also not immune to collective failures of metacognition. Millions of new scientific papers are published each year, with tens of thousands within my field of psychology alone. It would be impossible to read them all. And it would probably not be a good idea to try. In a wonderful book entitled *Ignorance*, the Columbia University neuroscientist Stuart Firestein argues persuasively that when facing a deluge of scientific papers, talking about what you don't know, and cultivating ignorance about what remains to be discovered, is a much more important skill than simply knowing the facts. I try to remind students in my lab that while knowledge is important, in science a critical skill is knowing enough to know what you don't know![10]

In fact, science as a whole may be becoming more and more aware of what it does and does not know. In a 2015 study, it was found that out of one hundred textbook findings in academic psychology, only thirty-nine could be successfully reproduced. A more recent replication of twenty-one high-profile studies published in *Science* and *Nature* found a slightly better return— 62 percent—but one that should still be concerning to social scientists wishing to build on the latest findings. This replication crisis takes on deeper meaning for students just embarking on their PhDs, who are often tasked with repeating a key study from another lab as a jumping-off point for new experiments. When apparently solid findings begin to disintegrate, months and even years can be wasted chasing down results that don't exist. But it is increasingly recognized that despite some findings in science being flaky and unreliable, many people already knew this! In bars outside scientific conferences and meetings, it's common to overhear someone saying, "Yeah I saw the latest finding from Bloggs et al., but I don't think it's real."[11]

In other words, scientists seem to have finely tuned bullshit detectors, but we have allowed their influence to be drowned out by a slavish adherence to the process of publishing traditional papers. The good news is that the system is slowly changing. Julia Rohrer at the Max Planck Institute for Human Development in Berlin has set out to make it easier for people to report a change of mind about their data, spearheading what she calls the Loss-of-Confidence Project. Researchers can now fill out a form explaining why they no longer trust the results of an earlier study that they themselves conducted. The logic is that the authors know their study best, and so are in the best place to critique it. Rohrer hopes the project will "put the self back into self-correction" in science and make things more transparent for other researchers trying to build on the findings.[12]

In addition, by loosening the constraints of traditional scientific publishing, researchers are becoming more used to sharing their data and code online, allowing others to probe and test their claims as a standard part of the peer-review process. By uploading a time-stamped document outlining the predictions and rationale for an experiment (known as preregistration), scientists can keep themselves honest and avoid making up stories to explain sets of flaky findings. There is also encouraging data that shows that when scientists own up about getting things wrong, the research community responds positively, seeing them as more collegiate and open rather than less competent.[13]

Another line of work is aiming to create "prediction markets" where researchers can bet on which findings they think will replicate. The Social Sciences Replication Project team set up a stock exchange, in which volunteers could buy or sell shares in each study under scrutiny, based on how reproducible they expected it to be. Each participant in the market started out with $100, and their final earnings were determined by how much they bet on the findings that turned out to replicate. Their choices of which studies to bet on enabled the researchers to precisely determine how much meta-knowledge the community had about its own work. Impressively, these markets do a very good job of predicting which studies will turn out to be robust and which won't. The traders were able to tap into features like the weakness of particular statistical results or small sample sizes—features that don't always preclude publication but raise quiet doubts in the heads of readers.[14]

This collective metacognition is likely to become more and more important as science moves forward. Einstein had a powerful metaphor for the paradox of scientific progress. If we imagine the sum total of scientific knowledge as a balloon, then as we blow up the balloon, the surface area expands and touches more and more of the unknown stuff outside of the balloon. The more knowledge we have, the more we do not know, and the more important it is to ask the right questions. Science is acutely reliant on self-awareness—it depends on individuals being able to judge the strength of evidence for a particular position and to communicate it to others. Stuart Firestein refers to this as "high quality" ignorance, to distinguish it from the low quality ignorance typified by knowing very little at all. He points out that "if ignorance. . . is what propels science, then it requires the same degree of care and thought that one accords data."

Creating a Self-Aware Society

We have seen that effective collaboration in situations ranging from a sports field to a court of law to a science lab depends on effective metacognition. But our social interactions are not restricted to handfuls of individuals in workplaces and institutions. Thanks to social media, each of us now has the power to share information and influence thousands, if not millions, of individuals. If our collective self-awareness is skewed, then the consequences may ripple through society.

Consider the role metacognition may play in the sharing of fake news on social media. If I see a news story that praises my favored political party, I may be both less likely to reflect on whether it is indeed real and more likely to just mindlessly forward it on. If others in my social network view my post, they may infer (incorrectly) that because I am usually a reliable source of information on social media, they don't need to worry about checking the source's accuracy—a case of faulty mindreading. Finally, due to the natural variation of trait-level metacognition in the population, just by chance some people will be less likely to second-guess whether they have the right knowledge on a topic and perhaps be more prone to developing extreme or inaccurate beliefs. What started out as a minor metacognitive blind spot may quickly snowball into the mindless sharing of bad information.[15]

To provide a direct test of the role of metacognition in people's beliefs about societal issues, we devised a version of our metacognition tasks that people in the United States could do over the Internet, and also asked volunteers to fill out a series of questions about their political views. From the questionnaire data, we could extract a set of numbers that told us both where people sat on the political spectrum (from liberal to conservative) and also how dogmatic or rigid they were in these views. For instance, dogmatic people tended to strongly agree with statements such as "My opinions are right and will stand the test of time." There was variability in dogmatism across a range of political views; it was possible to both be relatively centrist in one's politics and dogmatic. However, the most dogmatic individuals were found at both the left and right extremes of the political spectrum.

In two samples of over four hundred people, we found that one of the best predictors of holding dogmatic political views—believing that you are right and everyone else is wrong—was a lack of metacognitive sensitivity about simple perceptual decisions. Being dogmatic did not make you any worse at the task, but it did make you worse at knowing whether you were right or wrong in choosing which of two boxes contained more dots. This lack of metacognition also predicted the extent to which people would ignore new information and be unwilling to change their minds, especially when receiving information that indicated they were wrong about their original judgment. It's important to point out that this relationship was not specific to one or other political view. Dogmatic Republicans and dogmatic Democrats were equally likely to display poor metacognition.

Instead, people who were less self-aware were more likely to have dogmatic opinions about political issues of all kinds.[16]

In a subsequent study, we wanted to go a step further and ask whether subtle distortions in metacognition might also affect people's decisions to seek out information in the first place. As we have seen in the case of Jane studying for her exam, if our metacognition is poor we might think we know something when we don't and stop studying before we should. We wondered if a similar process might be at work in situations where people need to decide whether to seek out new information about topics such as politics and climate change. To investigate this, we adjusted our experiment slightly so that people were now also asked whether they would like to see the stimulus again in situations where they were unsure of the right answer. Doing so meant a small amount was deducted from their earnings in the experiment, but we made sure this cost was outweighed by the amount of points they received for getting a correct answer. We found that, on average, people decided to see the new information more when they were less confident, exactly as we would expect from a tight link between metacognition and information seeking. However, those people who were more dogmatic about political issues were also less likely to seek out new information, and their decisions to seek out new information were less informed by their confidence.[17]

These results indicate that poor metacognition may have quite general effects. By collecting confidence ratings on a simple dot-counting task, we could quantify people's capacity for metacognition in isolation from the kinds of emotional or social influences that often come along with decisions about hot-button issues such as politics. And yet, in all our studies so far, poor metacognition predicted whether people would hold extreme beliefs better than other, more traditional predictors in political science such as gender, education, or age.

This does not mean that specific factors are not playing a role in shaping how people reflect on and evaluate their beliefs. There may be some areas of knowledge that are particularly susceptible to metacognitive blind spots. The climate scientist Helen Fischer has quantified German citizens' self-awareness of various scientific topics, including climate change. People were given a series of statements—such as, "It is the father's gene that decides whether the baby is a boy or a girl," "Antibiotics kill viruses as well as bacteria," and "Carbon dioxide concentration in the atmosphere has increased more than 30 percent in the past 250 years"—and asked whether or not they are supported by science (the correct answers for these three questions would be yes, no, and yes). The volunteers also gave a confidence rating in their answers so that their metacognition could be quantified. People's metacognition tended to be quite good for general scientific knowledge. Even if they got some questions wrong, they tended to know that they were likely to be wrong, rating low confidence in their answers. But for climate change knowledge, metacognition was noticeably poor, even when controlling for differences in the accuracy of people's answers. It is not hard to see how such skewed metacognition might help drive the sharing of incorrect information on social media, causing ripples of fake news within a social network.[18]

Many conflicts in society arise from disagreements about fundamental cultural, political, and religious issues. These conflicts can become magnified when people are convinced that they are right and the other side is wrong. In contrast, what psychologists call intellectual humility—recognizing that we might be wrong and being open to corrective information—helps us diffuse these conflicts and bridge ideological gaps. Self-awareness is a key enabler of intellectual humility, and, when it is working as it should, it provides a critical check on our worldview.[19]

Thankfully, as we will see, there are ways of cultivating self-awareness and promoting it in our institutions and workplaces.

By understanding the factors that affect metacognitive prowess, we can capitalize on the power of self-awareness and avoid the pitfalls of metacognitive failure. For instance, by engaging in regular social interaction, team members can intuitively apply mindreading and metacognition to adapt to each other's communication style and avoid metacognitive mismatches when under pressure. For high-stakes decisions, such as whether or not to take on a new legal case or make a major business deal, submitting private, numerical estimates of confidence can increase the accuracy of a group's predictions. The process of communicating science or interrogating eyewitnesses can be structured to encourage a healthy degree of doubt and skepticism—just as with New Jersey's instructions to its judges. Those in positions of leadership—from lawyers to professors to football referees—can recognize that confidence is not always an indicator of competence and ensure that all voices, not just the loudest, are heard. Successful businesspeople—from Dalio to Bezos—know this. Innovative lawyers and scientists know this too.[20]

More broadly, collective self-awareness allows institutions and teams to change and innovate—to have autonomy over their futures, rather than mindlessly continuing on their current path. In the next chapter, we are going to return to the level of the individual to see that the same is also true of ourselves.

9

EXPLAINING OURSELVES

> It is the capacity to self-monitor, to subject the brain's patterns of reactions to yet another round (or two or three or seven rounds) of pattern discernment, that gives minds their breakthrough powers.
>
> —DANIEL DENNETT, *From Bacteria to Bach and Back*

When we set out to acquire a new skill such as playing tennis or learning to drive, we are often painfully aware of the minute details of our actions and why they might be going wrong. Metacognition is very useful in these early phases, as it tells us the potential reasons why we might have failed. It helps us diagnose why we hit the ball out of court and how to adjust our swing to make this less likely next time around. Rather than just registering the error and blindly trying something different, if we have a model of our performance, we can explain why we failed and know which part of the process to alter. Once a skill becomes well practiced, this kind of self-awareness is sometimes superfluous and often absent. In comparison to when we were student drivers, thoughts about how to drive are usually the last thing on our minds when we jump in the car to head off to work.[1]

The psychologist Sian Beilock aimed to quantify this relationship between skill and self-awareness in a landmark study of expert golfers. She recruited forty-eight Michigan State undergraduates, some of whom were intercollegiate golf stars and others who were golf novices. Each student was asked to make a series of "easy" 1.5 meter putts to a target marked on an indoor putting green. The golf team members showed superior putting ability, as expected, but also gave less detailed descriptions of the steps that they were going through during a particular putt, what Beilock called "expertise-induced amnesia." It seemed they were putting on autopilot and couldn't explain what they had just done. But when the experts were asked to use a novelty putter (one with an S-shaped bend and weights attached), they began to attend to their performance and gave just as detailed descriptions of what they were doing as the novices.

Beilock suggested that attention to our performance benefits novices in the earlier stages of learning but becomes counterproductive as movements become more practiced and routine. In support of this idea, she found that when expert golfers were instructed to monitor a series of beeps on a tape recorder while putting instead of attending to their swing, they became *more* accurate in their performance. Conversely, choking under pressure may be a case of the pilot interfering too much with the autopilot.[2]

This also suggests an explanation of why the best athletes don't always make the best coaches. To teach someone else how to swing a golf club, you first need to be able to explain how *you* swing a golf club. But if you have already become an expert golfer, then this is exactly the kind of knowledge we would expect you to have lost some time ago. Players and fans often want a former champion at the helm, as it's assumed that, having won titles and matches themselves, they must be able to impart the relevant know-how to others. And yet the history of sport is littered with examples of star athletes who faltered once they transitioned into coaching.

The fates of two different managers associated with my home football team, Manchester United, help illustrate this point. In the 1990s, Gary Neville was part of the famous "class of '92"—one of the most decorated teams in the club's history. But his transition to management, at the storied Spanish club Valencia, was less successful. During his short time in charge, Valencia were beaten 7–0 by rivals Barcelona, crashed out of the Champions League, and became embroiled in a relegation battle in their domestic league. In contrast, José Mourinho played fewer than one hundred games in the Portuguese league. Instead, he studied sports science, taught at schools, and worked his way up the coaching ladder before going on to become one of the most successful managers in the world, winning league titles in Portugal, England, Italy, and Spain before a brief (and less successful) stint with United in 2016. As the sports scientists Steven Rynne and Chris Cushion point out, "Coaches who did not have a career as a player were able to develop coaching skills in ways former champions did not have the time to engage with—because they were busy maximising their athletic performances." When you are focused on becoming a world-class player, the self-awareness needed to teach and coach the same skills in others may suffer.[3]

Taken to an extreme, if a skill is always learned through extensive practice, rather than explicit teaching, it is possible to get caught in a loop in which teaching is not just inefficient but impossible. One of the strangest examples of the cultural absence of metacognitive knowledge is the world of chicken sexing. To churn out eggs commercially, it is important to identify female chicks as early as possible to avoid resources being diverted to unproductive males. The snag is that differentiating features, such as feather color, only emerge at around five to six weeks of age. It used to be thought that accurate sexing before this point was impossible. This all changed in the 1920s, when Japanese farmers realized that the skill of early sexing could be taught via trial-and-error learning. They opened the Zen-Nippon Chick Sexing School, offering two-year courses on how to sort the chicks by becoming sensitive to almost imperceptible differences in anatomy (known as vents) and turning Japan into a hotbed of chick-sexing expertise. Japanese sexers became rapidly sought after in the United States, and in 1935 one visitor wowed agricultural students by sorting 1,400 chicks in an hour with 98 percent accuracy.

While a novice will start out by guessing, expert sexers are eventually able to "see" the sex by responding to subtle differences in the pattern of the vents. But despite their near-perfect skill levels, chick sexers' metacognition is generally poor. As the cognitive scientist Richard Horsey explains, "If you ask the expert chicken sexers themselves, they'll tell you that in many cases they have no idea how they make their decisions." Apprentice chick sexers must instead learn by watching their masters, gradually picking up signs of how to sort the birds. The best chick sexers in the world are like those star footballers who never made it as coaches; they can perform at the highest level, without being able to tell others how.[4]

Another example of skill without explainability is found in a strange neurological condition known as blindsight. It was first discovered in 1917 by George Riddoch, a medic in the Royal Army Medical Corps tasked with examining soldiers who had suffered gunshot wounds to the head. Soldiers with damage to the occipital lobe had cortical blindness—their eyes were still working, but regions of the visual system that received input from the eyes were damaged. However, during careful testing, Riddoch found that a few of the patients in his sample were able to detect moving objects in their otherwise "blind" field of view. Some residual processing capacity remained, but awareness was absent.

The Oxford psychologist Lawrence Weiskrantz followed up on Riddoch's work by intensively studying a patient known as "DB" at the National Hospital for Neurology in central London, an imposing redbrick building opposite our lab at UCL. DB's occipital cortex on the right side had been removed during surgery on a brain tumor, and he was cortically blind on the left side of space as a result. But when Weiskrantz asked DB to guess whether a visual stimulus was presented at location A or B, he could perform well above chance. Some information was getting into his brain, allowing him to make a series of correct guesses. DB was unable to explain how he was doing so well and was unaware he had access to any visual information. To him, it felt like he was just blind. Recent studies using modern anatomical tracing techniques have shown that blindsight is supported by information from the eyes traveling via a parallel and evolutionarily older pathway in the brain stem. This pathway appears able to make decisions about simple stimuli without the cortex getting involved.[5]

At first glance, blindsight seems like a disorder of vision. After all, it starts with the visual cortex at the back of the brain being damaged. And it is indeed the case that blindsight patients feel as though they are blind; they don't report having conscious visual experience of the information coming into their eyes. And yet, one of the hallmarks of blindsight is a lack of explainability, an inability for the normal machinery of self-awareness to gain access to the information being used to make guesses about the visual stimuli. As a result, blindsight patients' metacognition about their visual decisions is typically low or absent.[6]

These studies tell us two things. First, they reinforce the idea that self-awareness plays a central role in being able to teach things to others. If I cannot know how I am performing a task, I will make a poor coach. They also highlight how metacognition underpins our ability to explain what we are doing and why. In the remainder of this chapter, we are going to focus on this subtle but foundational role of self-awareness in constructing a narrative about our behavior—and, in turn, providing the bedrock for our societal notions of autonomy and responsibility.

The Interpreter

Imagine that it's a Saturday morning and you are trawling through your local supermarket. At the end of an aisle, a smiling shop assistant is standing behind a jam-tasting stall. You say hello and dip a plastic spoon into each of the two jars. The assistant asks you: Which one do you prefer? You think about it briefly, pointing to the jam on the left that tasted a bit like grapefruit. She gives you another taste of the chosen pot, and asks you to explain why you like it. You describe the balance of fruit and sweetness, and think to yourself that you might even pick up a jar to take home. You're just about to wander off to continue shopping when you're stopped. The assistant informs you that this isn't just another marketing ploy; it's actually a live psychology experiment, and with your consent your data will be analyzed as part of the study. She explains that, using a sleight of hand familiar to stage magicians, you were actually given the opposite jar to the one you chose when asked to explain your choice. Many other people in the study responded just as you did. They went on to enthusiastically justify liking a jam that directly contradicted their original choice, with only around one-third of the 180 participants detecting the switch.[7]

This study was carried out by Lars Hall, Petter Johansson, and their colleagues at Lund University in Sweden. While choices about jam may seem trivial, their results have been replicated in other settings, from judging the attractiveness of faces to justifying political beliefs. Even if you indicated a clear preference for environmental policies on a political survey, being told you in fact gave the opposite opinion a few minutes earlier is enough to prompt a robust round of contradictory self-justification. This phenomenon, known as choice blindness, reveals that we often construct a narrative to explain why we chose what we did, even if this narrative is partly or entirely fiction.[8]

It is even possible to hone in on the neural machinery that is involved in constructing narratives about our actions. In cases of severe epilepsy, a rare surgical operation is sometimes performed to separate the two hemispheres of the brain by cutting the large bundle of connections between them, known as the corpus callosum. Surprisingly, despite their brains being sliced in two, for most of these so-called split-brain patients the surgery is a success—the seizures are less likely to spread through the brain—and they do not feel noticeably different upon waking. But with careful laboratory testing, some peculiar aspects of the split-brain condition can be uncovered.

Michael Gazzaniga and Roger Sperry pioneered the study of split-brain syndrome in California in the 1960s. Gazzaniga developed a test that took advantage of how the eyes are wired up to the brain: the left half of our visual field goes into the right hemisphere, whereas the right half is processed by the left hemisphere. In intact brains, whatever information the left hemisphere receives is rapidly transferred to the right via the corpus callosum and vice versa (which is partly why the idea of left- and right-brained processing is a myth). But in split-brain patients, it is possible to send in a stimulus that remains sequestered in one hemisphere. When this is done, something remarkable happens. Because in most people the ability for language depends on neural machinery in the left hemisphere, a stimulus can be flashed on the left side of space (and processed by the right brain) and the patient will deny seeing anything. But because the right hemisphere can control the left hand, it is still possible for the patient to signal what he has seen by drawing a picture or pressing a button. This can lead to some odd phenomena. For instance, when the instruction "walk" was flashed to the right hemisphere, one patient immediately got up and left the room. When asked why, he responded that he felt like getting a drink. It seems that the left hemisphere was tasked with attempting to make sense of what the patient was doing, but without access to the true reason, which remained confined to the right hemisphere. Based on data such as this, Gazzaniga refers to the left hemisphere as the "interpreter."[9]

This construction of a self-narrative or running commentary about our behavior has close links with the neural machinery involved in metacognition—specifically, the cortical midline structures involved in self-reflection and autobiographical memory that we encountered in Chapter 3. In fact, the origins of the word narrative are from the Latin *narrare* (to tell), which comes from the Indo-European root *gnarus* (to know). Constructing a self-narrative shares many characteristics with the construction of self-knowledge. Patients with frontal lobe damage—the same patients who, as we saw in Chapter 4, have problems with metacognition—also often show strange forms of self-narrative, making up stories about why they are in the hospital. One patient who had an injury to the anterior communicating artery, running through the frontal lobe, confidently claimed that his hospital pajamas were only temporary and that he was soon planning to change into his work clothes. One way of making sense of these confabulations is that they are due to impaired self-monitoring of the products of memory, making it difficult to separate reality and imagination.[10]

These narrative illusions may even filter down to our sense of control or agency over our actions. In October 2016, Karen Penafiel, executive director of the American company National Elevator Industry, Inc., caused a stir by announcing that the "close door" feature in most elevators had not been operational for many years. Legislation in the early 1990s required elevator doors to remain open long enough for anyone with crutches or a wheelchair to get on board, and since then it's been impossible to make them close faster. This wouldn't have been much of a news story, aside from the fact that we still think these buttons make things happen. The UK's *Sun* newspaper screamed with the headline: "You Know the 'Close Door' Button in a Lift? It Turns Out They're FAKE." We feel like we have control over the door closing, and yet we don't. Our sense of agency is at odds with reality. The Harvard psychologist Daniel Wegner explained this phenomenon as follows:

> The feeling of consciously willing our actions . . . is not a direct readout of such scientifically verifiable will power. Rather, it is the result of a mental system whereby each of us estimates moment-to-moment the role that our minds play in our actions. If the empirical will were the measured causal influence of an automobile's engine on its speed, in other words, the phenomenal will might best be understood as the speedometer reading. And as many of us have tried to explain to at least one police officer, speedometer readings can be wrong.[11]

A clever experiment by Wegner and Thalia Wheatley set out to find the source of these illusions of agency. Two people sat at a single computer screen on which a series of small objects (such as a toy dinosaur or car) was displayed. They were both asked to place their hands on a flat board that was stuck on top of a regular computer mouse, allowing them to move the cursor together. Through headphones, the participants listened to separate audio tracks while continuing to move the mouse cursor around the screen. If a burst of music was heard, each participant was instructed to stop the cursor (a computerized version of musical chairs). They then rated how much control they felt over making the cursor stop on the screen.

Unbeknownst to the participant, however, there was a trick involved. The other person was a confederate, acting on the instructions of the experimenter. In some of the trials, the confederate heard instructions to move to a particular object on the screen, slow down, and stop. On these very same trials the participant heard irrelevant words through the headphones that highlighted the object they happened to be moving toward (for example, the word "swan" when the confederate was instructed to move toward the swan). Remarkably, on these trials, participants often felt more in control of making the cursor stop, and this responsibility went up when the object was recently primed. In other words, just thinking about a particular goal can cause us to feel in control of getting there, even when this feeling is an illusion. Other experiments have shown that people feel more agency when their actions are made more quickly and fluently, or when the consequences of their actions are made more predictable (such as a beep always following a button press). This mimics the case in most elevators: because the doors normally do close shortly after we press the "close door" button, we start to feel a sense of control over making this happen.[12]

Constructing Autonomy

As suggested by the fact that the left (rather than the right) hemisphere plays the role of interpreter in Gazzaniga's experiments, another powerful influence on self-narratives is language. From an early age, children begin to talk to themselves, first out loud, and then in their heads. Language—whether spoken or internalized—provides us with a rich set of tools for thinking that allows the formation of recursive beliefs (for instance: I think that I am getting ill, but it could be that I am stressed about work). This recursive aspect of language supercharges our metacognition, allowing us to create, on the fly, brand new thoughts about ourselves.[13]

But there is a problem with this arrangement. It can mean that what we think we are doing (our self-narratives) can begin to slowly drift away from what we are actually doing (our behavior). Even without a split brain, I might construct a narrative about my life that is subtly at odds with reality. For instance, I might aspire to get up early to write, head in to the lab to run a groundbreaking experiment, then return home to play with my son and have dinner with my wife, all before taking a long weekend to go sailing. This is an appealing narrative. But the reality is that I sometimes oversleep, spend most of the day answering email, and become grumpy in the evening due to lack of progress on all these fronts, leading me to skip weekends away to write. And so on.

Another way of appreciating this point is that our narratives need to be reasonably accurate to make sense of our lives, even if they may in part be based on confabulation. When the self-narrative drifts too far from reality, it becomes impossible to hold onto. If I had a narrative that I was an Olympic-level sailor, then I would need to construct a more extensive set of (false) beliefs about why I haven't been picked for the team (perhaps they haven't spotted me yet) or why I'm not winning races at my local club (perhaps I always pick the slow boat). This is a feature of the delusions associated with schizophrenia. But as long as our narratives broadly cohere with the facts, they become a useful shorthand for our aspirations, hopes, and dreams.[14]

A higher-level narrative about what we are doing and why can even provide the scaffold for our notions of autonomy and responsibility. The philosopher Harry Frankfurt suggested that humans have at least two levels of desires. Self-knowledge about our preferences is a higher-order desire that either endorses or goes against our first-order motives, similar to the notion of confidence about value-based decisions we encountered in Chapter 7. Frankfurt proposed that when our second-order and first-order desires match up, we experience heightened autonomy and free will. He uses the example of someone addicted to drugs who wants to stop. They want to not want the drugs, so their higher-order desire (wanting to give up) is at odds with their first-order desire (wanting the drugs). We intuitively would sympathize with such a person as struggling with themselves and perhaps think that they are making the choice to take drugs less of their own free will than another person who enthusiastically endorses their own drug taking.[15]

When second-order and first-order desires roughly line up—when our narratives are accurate—we end up wanting what we choose and choosing what we want. Effective metacognition about our wants and desires allows us to take steps to ensure that such matches are more likely to happen. For instance, while writing this chapter, I have already tried to check Twitter a couple of times, but my browser's blocking software stops me from being sidetracked by mindless scrolling. My higher-order desire is to get the book written, but I recognize that I have a conflicting first-order desire to check social media. I can therefore anticipate that I might end up opening Twitter and take advance steps (installing blocking software) to ensure that my actions stay true to my higher-order desires—that I end up wanting what I choose.[16]

The upshot of all this is an intimate link between self-knowledge and autonomy. As the author Al Pittampalli sums up in his book *Persuadable*, "The lesson here is this: Making the choice that matches your interests and values at the highest level of reflection, regardless of external influence and norms, is the true mark of self-determination. This is what's known as autonomy."[17] The idea of autonomy as a match between higher-order and first-order desires can seem lofty and philosophical. But it has some profound consequences. First, it suggests that our feeling of being in charge of our lives is a construction—a narrative that is built up at a metacognitive level from multiple sources. Second, it suggests that a capacity for metacognition may be important for determining whether we are to blame for our actions.

We can appreciate this close relationship between self-awareness and autonomy by examining cases in which the idea of responsibility is not only taken seriously but also clearly defined, as in the field of criminal law. A central tenet of the Western legal system is the concept of mens rea, or guilty mind. In the United States, the definition of mens rea has been somewhat standardized with the introduction of the Model Penal Code (MPC), which was developed in the 1960s to help streamline this complex aspect of the law. The MPC helpfully defines four distinct levels of culpability or blame:

- Purposely—one's *conscious* object is to engage in conduct of that nature

- Knowingly—one has *awareness* that one's conduct will cause such a result

- Recklessly—one has *conscious disregard* of a substantial and unjustifiable risk

- Negligently—one *should be aware* of a substantial and unjustifiable risk

I have highlighted the terms related to self-awareness in italics—which strikingly appear in every definition. It's clear that the extent to which we have awareness of our actions is central to legal notions of responsibility and blame. If we are unaware of what we are doing, then we may sometimes be excused even of the most serious of crimes, or at the very least would only be found negligent.

A tragic example is provided by a case of night terrors. In the summer of 2008, Brian Thomas was on holiday with his wife in Aberporth, in western Wales, when he had a vivid nightmare. He recalled thinking he was fighting off an intruder in their caravan, perhaps one of the kids who had been disturbing his sleep by revving motorbikes outside. In reality, he was gradually strangling his wife to death. He awoke from his nightmare into a living nightmare, and made a 999 call to tell the operator that he was stunned and horrified by what had happened, and entirely unaware of what he had done.

Crimes committed during sleep are rare, thankfully for both their perpetrators and society at large. Yet they provide a striking example of the potential to carry out complex actions while remaining unaware of what we are doing. In Brian Thomas's case, expert witnesses agreed that he suffered from a sleep disorder known as pavor nocturnus, or night terrors, which affects around 1 percent of adults and 6 percent of children. The court was persuaded that his sleep disorder amounted to "automatism," a comprehensive defense under UK law that denies even the lowest degree of mens rea. After a short trial, the prosecution simply withdrew their case.[18]

This pivotal role played by self-awareness in Western legal systems seems sensible. After all, we have already seen that constructing a narrative about our behavior is central to personal autonomy. But it also presents a conundrum. Metacognition is fragile and prone to illusions, and can be led astray by the influence of others. If our sense of agency is just a construction, created on the fly, then how can we hold people responsible for anything? How can we reconcile the tension between metacognition as an imperfect monitor and explainer of behavior with its critical role in signaling responsibility for our actions?

I think there are two responses to this challenge. The first is that we needn't worry, because our sense of agency over our actions is accurate enough for most purposes. Think of the last time that you excused a mistake with the comment, "Sorry, I just wasn't thinking." I suspect that your friends and family took you at your word, rather than indignantly complaining: "How can you possibly know? Introspection is a mental fiction!" And in most cases, they would be right to trust in your self-knowledge. If you really were "not thinking" when you forgot to meet a friend for lunch, then it is useful for others to know that it was not your intention to miss the appointment and that you can be counted on to do better in future. In many, even most, cases, metacognition works seamlessly, and we *do* have an accurate sense of what we are doing and why we are doing it. We should be in awe that the brain can do this at all.

The second response is to recognize that self-awareness is only a useful marker of responsibility in a culture that agrees that this is the case. Our notions of autonomy, like our legal systems, are formed out of social exchange. Responsibility, then, is rather like money. Money has value only because we collectively agree that it does. Similarly, because we collectively agree that self-awareness is a useful marker of a particular mode of decision-making, it becomes central to our conception of autonomy. And, just like money, autonomy and responsibility are ultimately creations of the human mind—creations that depend on our ability to build narratives about ourselves and each other. As with money, we can recognize this constructed version of responsibility while at the same time enjoying having it and using it.

A deeper implication of this tight link between self-awareness and responsibility is that if the former suffers, the latter might also become weakened. Such questions are becoming increasingly urgent in an aging population, which may suffer from diseases such as dementia that attack the neural foundations of self-awareness. Western democracies are grappling with how to strike the balance between preserving autonomy and providing compassionate support in these cases. Recently enacted laws, such as the UK's Mental Capacity Act, codify when the state should take charge of a person's affairs when they are no longer able to make decisions for themselves due to a psychiatric or neurological disorder. Other proposals, such as that put forward in the United Nations Convention on the Rights of Persons with Disabilities, hold that liberty should be maintained at all costs. At the heart of this battle is a fight for our concept of autonomy: Under what circumstances can and should our friends, family, or government step in to make decisions on our behalf?[19]

While the law in this area is complex, it is telling that one commonly cited factor in capacity cases is that the patient lacked insight or self-awareness. This implies that systematic changes in our own self-awareness may affect our chances of continuing to create a narrative and hold autonomy over our lives. The potential for these changes might be more widespread than we think, arriving in the form of new technology, drugs, or social structures. In the final two chapters of this book, we are going to explore what the future holds for human self-awareness—from the need to begin to coordinate and collaborate with intelligent machines, to the promise of technology for supercharging our ability to know ourselves.

10

SELF-AWARENESS IN THE AGE OF MACHINES

> As technology progresses, an ever more intimate mix of
> human and machine takes shape. You're hungry; Yelp
> suggests some good restaurants. You pick one; GPS gives
> you directions. You drive; car electronics does the low-level
> control. We are all cyborgs already.
>
> —PEDRO DOMINGOS, *The Master Algorithm*

> From the entirely objective point of view of information and
> computer theories, all scientific knowledge of dynamic
> systems is knowledge of the aspect that is machine-like.
> Nevertheless, the questions are still being asked: Can a
> machine know it is a machine? Has a machine an internal
> self-awareness?
>
> —W. ROSS ASHBY, *Mechanisms of Intelligence*

In June 2009, an Air France flight from Rio de Janeiro to Paris
disappeared into the Atlantic Ocean. The three pilots were flying an Airbus A330—one of the most advanced aircraft in the
world, laden with autopilot and safety mechanisms and notoriously difficult to crash. As night fell during the ocean crossing,
the pilots noticed a storm on the flight path. Storms are usually

straightforward for modern airliners to handle, but on this occasion ice in the clouds caused a sensor on the plane to seize up and stop functioning, leading the autopilot to disconnect and requiring the pilots to take control. Pierre-Cédric Bonin, the inexperienced copilot, tried to take over manual control, but he began flying unsteadily and started to climb. The thin air made the aircraft stall and lose altitude.

Even at this stage, there was little danger. All the pilots needed to do to fix the problem was to level out the plane and regain airspeed. It's the kind of basic maneuverer that is taught to novice pilots in their first few hours of flying lessons. But Bonin kept on climbing, with disastrous consequences. The investigation report found that, despite the pilots having spent many hours in the cockpit of the A330, most of these hours had been spent monitoring the autopilot, rather than manual flying. They were unwilling to believe that all this automation would let them make the series of errors that ended in disaster.

The moral of this tragic story is that sometimes offloading control to automation can be dangerous. We have already seen that a well-learned skill such as a golf swing can become automatic to the point at which the need to think about what we are doing becomes less necessary and even detrimental for performing well. But this kind of automation is all within the same brain—our own. When things go awry—when we hook a golf ball, fluff a tennis backhand, or put the car in the wrong gear—we are usually jolted back into awareness of what we are doing. The picture is very different when a machine has taken control. Paradoxically, as automation becomes more and more sophisticated, human operators become less and less relevant. Complacency sets in, and skills may decline.

Such worries about the consequences of offloading to technology are not new. Socrates tells the mythical story of Egyptian god Theuth, who is said to have discovered writing. When

Theuth offered the gift to Thamus, king of Egypt, the king was not impressed and worried that it would herald the downfall of human memory, introducing a pandemic of forgetfulness. He complained that people who used it "will appear to be omniscient and will generally know nothing; they will be tiresome company, having the show of wisdom without the reality."[1]

Thankfully, these worries about writing and reading did not come to pass. But AI and machine learning may prove different. The intellectual boosts provided by writing, or the printing press, or even the computer, Internet, or smartphone, are the result of transparent systems. They respond lawfully to our requests: We set up a particular printing block, and this systematically reproduces the same book, time and time again. Each time I press the return key when writing on my laptop, I get a new line. The intersection of machine learning and automation is different. It is often nontransparent; we not only do not always know how it is working, but we also do not know whether it will continue to work in the same way tomorrow, the next day, or the day after that. In one sense, machine learning systems have minds of their own, but they are minds that are not currently able to explain how they are solving a particular problem. The remarkable developments in artificial intelligence have not yet been accompanied by comparable developments in artificial self-awareness.

In fact, as technology gets smarter, the relevance of *our* self-awareness might also diminish. A powerful combination of data and machine learning may end up knowing what we want or need better than we know ourselves. The Amazon and Netflix recommendation systems offer up the next movie to watch; dating algorithms take on the job of finding our perfect match; virtual assistants book hair appointments before we are aware that we need them; online personal shoppers send us clothes that we didn't even know we wanted.

As human consumers in such a world, we may no longer need to know how we are solving problems or making decisions, because these tasks have become outsourced to AI assistants. We may end up with only weak metacognitive contact with our machine assistants—contact that is too thin for us to intervene when they might be doing things they were not designed to do, or, if we are alerted, we may find it is too late to do anything about it. Machines would not need to explain how they are solving problems or making decisions either, because they had no need to do so in the first place. The outsourcing of intelligence could lead to meaningful self-awareness gradually fading into the background of a technology-dependent society.

So what? we might say. A radical futurist may see this melding of mind and machine as the next logical step in human evolution, with changes to how we think and feel being a small price to pay for such technological advances. But I think we need to be careful here. We have already seen in this book that any kind of mind—silicon or biological—is likely to need a degree of metacognition in order to solve scientific and political problems on a grand scale.

I see two broad solutions to this problem:

- Seek to engineer a form of self-awareness into machines (but risk losing our own autonomy in the process).

- Ensure that when interfacing with future intelligent machines, we do so in a way that harnesses rather than diminishes human self-awareness.

Let's take a look at each of these possibilities.

Self-Aware Machines

Ever since Alan Turing devised the blueprints for the first universal computer in 1937, our position as owners of uniquely intelligent minds has looked increasingly precarious. Artificial neural networks can now recognize faces and objects at superhuman speed, fly planes or pilot spaceships, make medical and financial decisions, and master traditionally human feats of intellect and ingenuity such as chess and computer games. The field of machine learning is now so vast and fast-moving that I won't attempt a survey of the state of the art and instead refer readers to excellent recent books on the topic in the endnote. But let's try to extract out a few key principles by considering how machine learning relates to some of the building blocks of metacognition we encountered in Part I.[2]

A useful starting point is to look at the components needed for a robot to begin to perceive its environment. From there, we can see what extra components we might need to create a form of machine self-awareness. Let's call our robot Val, after the cyberneticist Valentino Braitenberg, whose wonderful book *Vehicles* was one of the inspirations for this chapter. Val is pictured on page 194. She is a toy car with a camera on the front and motors on the wheels. There are two lights positioned in front of her: a blue light to the right, and a green light to the left. But so far, Val is nothing more than a camera. Her eyes work, but there is no one home. To allow Val to start seeing what is out there, we need to give her a brain.[3]

An artificial neural network is a piece of computer software that takes an input, such as a digital image, and feeds the information through a set of simulated layers of "neurons." In fact, each neuron is very simple: it computes a weighted sum of the inputs from neurons in the layers below, and passes the result

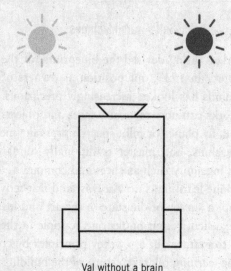

Val without a brain

through a nonlinear function to generate its level of "activation." This value is then passed on to the next layer, and so on. The clever thing about neural networks is that the weights connecting each layer can be adjusted, through trial and error, to begin to classify the inputs that it is fed. For instance, if you present a neural network with a series of images of cats and dogs and ask it to respond "cat" or "dog," you can tell the network each time it is right and each time it is wrong. This is known as supervised learning—the human is supervising the neural network to help it get better at the task of classifying cats and dogs. Over time, the weights between the layers are adjusted so that the network gets better and better at giving the right answer all by itself.[4]

Neural networks have a long history in artificial intelligence research, but the first networks were considered too simple to compute anything useful. That changed in the 1980s and 1990s with the advent of more computing power, more data, and clever ways of training the networks. Today's artificial neural

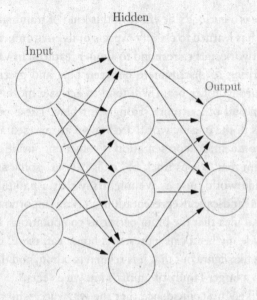

A diagram of an artificial neural network *(Glosser.ca)*

networks can classify thousands of images with superhuman performance. One particularly efficient approach is to use deep networks, which have multiple layers, just like the visual system of the human brain.[5]

To allow Val to start perceiving the world, then, we could hook up all the pixels from her digital camera to the input layer of a deep neural network. We could switch on each of the blue and green lights in turn, and train the output layers of the network to categorize the light as either blue or green. By providing corrective feedback (just as we might do to an infant who is learning the names of things in the world), the weights between the layers of Val's network would gradually get adjusted so that she is reliably perceiving the blue light as blue and the green light as green.

There is a curious side effect of this kind of training. Once a network has learned to classify aspects of its environment, parts of the network start to respond to similar features. In Val's case, after learning to discriminate between blue and green lights, some neurons in her brain will tend to activate more strongly for green, and others more strongly for blue. These patterned responses to the outside world are known as representations of the environment. A representation is something inside a cognitive system that "keeps track of" or is "about" some aspect of the outside world, just as we might say that a painting "represents" a landscape. Representations play an important role in the general idea that the brain performs computations. If something inside my head can represent a house cat, then I can also do things like figure out that it is related to a lion, and that both belong to a larger family of animals known as cats.[6]

It is likely no coincidence that the way successful artificial image-classification networks are wired is similar to the hierarchical organization of the human brain. Lower layers contain neurons that handle only small parts of the image and keep track of features such as the orientation of lines or the difference between light and shade. Higher layers contain neurons that process the entire image and represent things about the object in the image (such as whether it contains features typical of a cat or a dog). Computational neuroscientists have shown that exactly this kind of progression—from computing local features to representing more global properties—can be found in the ventral visual stream of human and monkey brains.[7]

Scaled-up versions of this kind of architecture can be very powerful indeed. By combining artificial neural networks with reinforcement learning, the London-based technology company DeepMind has trained algorithms to solve a wide range of board and video games, all without being instructed about the rules in advance. In March 2016, its flagship algorithm, AlphaGo, beat Lee Sedol, the world champion at the board game Go and one of the greatest players of all time. In Go, players take turns placing their stones on intersections of a nineteen-by-nineteen grid, with the objective of encircling or capturing the other player's stones. Compared to chess, the number of board positions is vast, outstripping the estimated number of atoms in the universe. But by playing against itself millions of times, and updating its predictions about valuable moves based on whether it won or lost, AlphaGo could achieve superhuman skill at a game that is considered so artful that it was once one of four essential skills that Chinese aristocrats were expected to master.[8]

These kinds of neural networks rely on supervised learning. They have to learn whether they are right or wrong by training on a series of examples. After being trained, they acquire rich representations of their environment and reward functions that tell them what is valuable. These algorithms can be stunningly powerful, intelligent, and resourceful, but they have limited self-awareness of what they do or do not know. It is also unlikely that self-awareness will simply emerge as a by-product of designing ever more intelligent machines. As we have seen, good ability does not necessarily lead to good metacognition. You might be performing expertly at a task (recall the chick sexers) and yet have no self-awareness of what you are doing. Instead, this kind of AI is likely to become more and more intelligent in particular domains, while perhaps remaining no more self-aware than a pocket calculator.

We can make this discussion more precise by considering what building blocks machines would need to become self-aware. Many of these components are things we have already encountered in Part I, such as the ability to track uncertainty and self-monitor actions. Machines often do not have these second-order capabilities—partly because, in most cases in which the problem is clearly specified, they do not need to. William James, the grandfather of modern psychology, anticipated this idea when he mused that a machine with a "clock-work

whose structure fatally determines it to a certain rate of speed"
would be unlikely to make errors in the first place, let alone
need to correct them. He contrasted this against the effortless
self-awareness of the human mind: "If the brain be out of order
and the man says 'Twice four are two,' instead of 'Twice four
are eight'. . . instantly there arises a consciousness of error."[9]

The machines in James's day were simple enough that errors
were the exception rather than the rule. This is no longer the
case. In fact, a key problem with modern machine learning techniques is that they are often overconfident in the real world;
they think they know the answer when they would be better off
hedging their bets. This poses a serious problem for operating
AI devices in novel environments—for instance, the software
installed in self-driving cars can be fooled by inputs it has not
encountered before or different lighting conditions, potentially
leading to accidents.[10]

Another problem is that, once a neural network is trained,
it is hard to know why it is doing what it is doing. As we have
seen, modern AI is not usually set up with the goal of self-
explanation. The philosopher Andy Clark and psychologist
Annette Karmiloff-Smith anticipated this idea in a landmark
article in 1993. They suggested that artificial neural networks
are unable to monitor what they are doing precisely because
their knowledge remains stored "in" the system, in the weights
connecting each layer.

They gave the example of training an artificial neural network to predict whether individuals would default on their
loans (something that was hypothetical in 1993 but is now a
routine aspect of machine learning at major banks). The network can draw on a huge database of information (such as
postal address or income level) about people in a particular
country (such as the UK) to refine its predictions about who will
and will not prove likely to default. Clark and Karmiloff-Smith
asked whether, after all this learning, the network would be able

to communicate what it had learned to a new network that the bank wanted to set up in Australia. In other words, would it know what it had learned and be able to use that self-knowledge to teach others? They concluded it would not: "What the system needs, in order to be able to tell the Australian system anything useful, is some more abstract and transportable knowledge concerning relevant factors in loan assessment. . . . The original network gets by without such explicit abstractions, since it only develops the minimal representations needed to succeed in the version of the task for which it was trained." In other words, Clark and Karmiloff-Smith were suggesting that, even with generous computing resources and ample data, neural networks solving complex problems are unlikely to become self-aware of what they know. Knowledge within these networks remains buried in a changing pattern of weights between the layers—it is knowledge "in" the network, rather than "for" the network. They went on to suggest that, in self-aware minds, a process of "representational redescription" also occurs (at least for some types of learning), allowing us not only to perceive and categorize a cat or a dog, but also to know that we are perceiving a cat as a cat and a dog as a dog.[11]

This all sounds quite abstract and theoretical. But the process of creating meta-representations is actually relatively simple. Just as we can have neural networks that take in information from the outside world, transform it, and spit out an answer, we can have metacognitive networks that model how other neural networks are operating.

One of the first attempts at creating artificial metacognition was made by Nicholas Yeung, Jonathan Cohen, and Matthew Botvinick in a landmark paper published in *Psychological Review* in 2004. They trained a neural network to solve a Stroop task (stating the color of a word rather than reading the word out loud), similar to the one we encountered earlier in the experiments on typewriting. The researchers then added a simple metacognitive network to monitor what was happening in their first network. In fact, this second network was so simple that it was just a single unit, or neuron, receiving inputs from the main network. It calculated how much "conflict" there was between the two responses: if both the word and the color were competing for control over the response, this would indicate that the judgment was difficult, and that an error might occur, and it would be wise to slow down. The simple addition of this layer was able to account for an impressive array of data on how humans detect their own errors on similar tasks.[12]

Other work has focused on attempting to build computer simulations of neurological conditions, such as blindsight, in which metacognition is impaired but lower-level performance is maintained. Axel Cleeremans's team has attempted to mimic this phenomenon by training an artificial neural network to discriminate among different stimulus locations. They then built a second neural network that tracked both the inputs and outputs of the first network. This metacognitive network was trained to gamble on whether the first network would get the answer correct. By artificially damaging the connections between the two networks, the team could mimic blindsight in their computer simulation. The first network remained able to correctly select the location, but the system's metacognitive sensitivity—the match between confidence and performance—was abolished, just as in the patients. In other experiments looking at how metacognitive knowledge emerges over time, Cleeremans's team found that the simulation's self-awareness initially lagged behind its performance; it was as if the system could initially perform the task intuitively, without any aware-ness of what it was doing.[13]

Various ingenious solutions are now being pursued to build a sense of confidence into AI, creating "introspective" robots that know whether they are likely to be right before they make a decision, rather than after the fact. One promising approach, known as dropout, runs multiple copies of the network, each with a slightly different architecture. The range of predictions that the copies make provides a useful proxy for how uncertain the network should be about its decision. In another version of these algorithms, autonomous drones were trained to navigate themselves around a cluttered environment—just as a parcel-delivery drone would need to do if operating around the streets and skyscrapers of Manhattan. The researchers trained a second neural network within the drone to detect the likelihood of crashes during test flights. When their souped-up introspective drone was released in a dense forest, it was able to bail out of navigation decisions that it had predicted would lead to crashes. By explicitly baking this kind of architecture into machines, we may be able to endow them with the same metacognitive building blocks that we saw are prevalent in animals and human infants. No one is yet suggesting that the drone is self-aware in a conscious sense. But by being able to reliably predict its own errors, it has gained a critical building block for metacognition.[14]

There may be other side benefits of building metacognition into machines, beyond rectifying their tendency for overconfidence. Think back to the start of the book, where we encountered our student Jane studying for an upcoming exam. In making decisions about when and where to study, Jane is likely drawing on abstract knowledge she has built up about herself over many instances, regardless of the subject she was studying. The neural machinery for creating abstract knowledge about ourselves is similar to the kind of machinery needed for creating abstract knowledge about how things work more generally. For instance, when going to a foreign country, you might not know how to operate the metro system, but you expect, based on experience in other cities, that there will be similar components such as ticket machines, tickets, and barriers. Leveraging this shared knowledge makes learning the new system much faster.

And such abstractions are exactly what Andy Clark and Annette Karmiloff-Smith recognized that we need to build into neural networks to allow them to know what they know and what they don't know.[15]

Knowledge about ourselves is some of the most transferable and abstract knowledge of all. After all, "I" am a constant feature across all situations in which I have to learn. Beliefs about my personality, skills, and abilities help me figure out whether I will be the kind of person who will be able to learn a new language, play an unfamiliar racket sport, or make friends easily. These abstract facts about ourselves live at the top of our metacognitive models, and, because of their role in shaping how the rest of the mind works, they exert a powerful force on how we live our lives. It is likely that similar abstract self-beliefs will prove useful in guiding autonomous robots toward tasks that fit their niche—for instance, to allow a drone to know that it should seek out more parcel-delivery jobs rather than try to vacuum the floor.

Let's imagine what a future may look like in which we are surrounded by metacognitive machines. Self-driving cars could be engineered to glow gently in different colors, depending on how confident they were that they knew what to do next— perhaps a blue glow for when they are confident and yellow for when they are uncertain. These signals could be used by their human operators to take control in situations of high uncertainty and increase the humans' trust that the car *did* know what it was doing at all other times. Even more intriguing is the idea that these machines could share metacognitive information with each other, just as self-awareness comes into its own when humans begin to collaborate and interact. Imagine two autonomous cars approaching an intersection, each signaling to turn in different directions. If both have a healthy blue glow, then they can proceed, safe in the knowledge that the other car has a good idea of what is happening. But if one or both of them begins to glow yellow, it would be wise to slow down and proceed with caution, just as we would do if a driver on the other side of the intersection didn't seem to know what our intentions were.

Intriguingly, this kind of exchange would itself be interactive and dynamic, such that if one car began to glow yellow, it could lead others to drop their confidence too. Every car at the intersection would begin to hesitate until it was safe to proceed. The sharing of metacognitive information between machines is still a long way from endowing them with a full-blown theory of mind. But it may provide the kind of minimal metacognitive machinery that is needed to manage interactions between human-machine and machine-machine teams. More elaborate versions of these algorithms—for instance, versions that know *why* they are uncertain (Is it the change in lighting? A new vehicle it's never encountered before?)—may also begin to approach a form of narrative explanation about why errors were made.

That is our first scenario: building minimal forms of artificial metacognition and self-awareness into machines. This research is already well underway. But there is also a more ambitious alternative: augmenting machines with the biology of human self-awareness.

Know Thy Robot

Imagine that when we step into a self-driving car of the future, we simply hook it up to a brain-computer interface while we drive it around the block a few times. The signals streaming back from the car while we drive gradually lead to changes in neural representations in the PFC just as they have already been shaped by a variety of other tools we use. We could then let the car take over; there would be nothing left for "us" to do in terms of routine driving. Critically, however, the brain-computer interface would ensure we have strong, rather than weak, metacognitive contact with the car. Just as we have limited awareness of the moment-to-moment adjustments of our actions, our vehicle would go places on our behalf. But if things were to go awry, we would naturally become aware of this—much as we might become aware of stumbling on a stationary escalator or hitting a poor tennis shot.

This scenario seems far-fetched. But there is nothing in principle to stop us from hooking up our self-awareness to other devices. Thanks to the plasticity of neural circuits, we already know it is possible for the brain to adopt external devices as if they were new senses or limbs. One of the critical developments in brain-computer interfaces occurred in the early 1980s. Apostolos Georgopoulos, a neuroscientist then at Johns Hopkins University, was recording the activity of multiple neurons in the monkey motor cortex. He found that when the monkey made arm movements in different directions, each cell had a particular direction for which its rate of firing was highest. When the firing of the entire population of cells, each with a different preferred direction, was examined, the vector sum of firing rates could predict with substantial accuracy where the monkey's hand actually went. It was not long before other labs were able to decode these population codes and show that monkeys could be trained to control robot arms by modulating patterns of neural activity.[16]

Matt Nagle, a tetraplegic left unable to move after being attacked and stabbed, was one of the first patients to receive the benefit of this technology in 2002. After receiving an implant made by a commercial company, Cyberkinetics, he learned to move a computer cursor and change TV channels by thought alone. Companies such as Elon Musk's Neuralink have recently promised to accelerate the development of such technologies by developing surgical robots that can integrate the implant with neural tissue in ways that would be near impossible for human surgeons. We might think that undergoing neurosurgery is a step too far to be able to control our AI devices. But other companies are harnessing noninvasive brain-scanning devices such as EEG, which, when combined with machine learning, may allow similarly precise control of external technology.[17]

Most current research on brain-computer interfaces is seeking to find ways of harnessing brain activity to control external technology. But there seems no principled reason why it would not also be possible for the brain to monitor autonomous devices rather than control them directly. Remember that metacognition has a wide purview. If we can interface autonomous technology with the brain, it is likely that we will be able to monitor it using the same neural machinery that supports self-awareness of other cognitive processes. In these cases, our brain-computer interface would be tapping into a higher level of the system: the level of metacognition and self-awareness, rather than perception and motor control.[18]

By designing our partnership with technology to take advantage of our natural aptitude for self-awareness, we can ensure humans stay in the loop. Progress in AI will provide new and rich sources of raw material to incorporate into our metacognitive models. If the pilots of the doomed Air France Airbus had had such a natural awareness of what their autopilot was doing, the act of stepping in and taking over control may have not been such a jarring and nerve-racking event.

It may not matter that we don't understand how these machines work, as long as they are well interfaced with metacognition. Only a small number of biologists understand in detail how the eye works. And yet, as humble users of eyes, we can instantly recognize when an image may be out of focus or when we need the help of reading glasses. Few people understand the complex biomechanics of how muscles produce the movements of our arms, and yet we can recognize when we have hit a tennis serve or golf swing poorly and need to go back to have more coaching. In exactly the same way, the machines of the future may be monitored by our own biological machinery for self-awareness, without us needing an instruction manual to work out how to do so.

What Kind of World Do We Want?

Which route we pursue depends on which world we want to live in. Do we want to share our world with self-aware machines? Or would we rather our AIs remain smart and un-self-aware, helping us to augment our natural cognitive abilities?

One concern with the first route is a moral one. Given the human tendency to ascribe moral responsibility to agents that possess self-awareness, enabling machines with even the first building blocks of metacognition may quickly raise difficult questions about the rights and responsibilities of our robot collaborators. For now, though, the richer metacognitive networks being pursued by AI researchers remain distinct from the flexible architecture of human self-awareness. Prototype algorithms for metacognition—such as a drone predicting that it might be about to crash—are just as un-self-aware as the kind of regular neural networks that allow Facebook and Google to classify images and Val to navigate her toy world.

Instead, current versions of artificial metacognition are quite narrow, learning to monitor performance on one particular task, such as classifying visual images. In contrast, as we have seen, human self-awareness is a flexible resource and can be applied to evaluate a whole range of thoughts, feelings, and behaviors. Developing domain-specific metacognitive capacity in machines, then, is unlikely to approach the kind of self-awareness that we associate with human autonomy.[19]

A second reason for the brittleness of most computer simulations of metacognition is that they mimic implicit, or "model-free," ways of computing confidence and uncertainty, but do not actively model what the system is doing. In this sense, AI metacognition has started to incorporate the unconscious building blocks of uncertainty and error monitoring that we encountered in Part I, but not the kind of explicit metacognition that, as we saw, emerges late in child development and is linked to our awareness of other minds. It may be that, if a second-order, Rylean view of self-awareness is correct, then humanlike self-awareness will emerge only if and when AI achieves a capacity for fully-fledged theory of mind.

But we should also not ignore what the future holds for our own self-awareness. Paradoxically, the neuroscience of meta-cognition tells us that by melding with AI (the second route) we may retain more autonomy and explainability than if we continue on the path toward creating clever but unconscious machines. This paradox is highlighted in current debates about explainable AI. Solutions here often focus on providing read-outs or intuitive visualizations of the inner workings of the black box. The idea is that if we can analyze the workings of the machine, we will have a better idea of why it is making partic-ular decisions. This might be useful for simple systems. But for complex problems it is unlikely to be helpful. It would be like providing an fMRI scan (or worse, a high-resolution map of prefrontal cortical cell firing) to explain why I made a particular decision about which sandwich to have for lunch. This would not be an explanation in the usual sense of the term and would not be one that would be recognized in a court of law. Instead, for better or worse, humans effortlessly lean on self-awareness to explain to each other why we did what we did, and a formal cross-examination of such explanations forms the basis of our sense of autonomy and responsibility.[20]

In practice, I suspect a blend of both approaches will emerge. Machines will gain domain-specific abilities to track uncertainty and monitor their actions, allowing them to effectively collabo-rate and build trust with each other and their human operators. By retaining humans in the loop, we can leverage our unique capacity for self-narrative to account for why our machines did what they did in the language of human explanation. I suspect it was this metacognitive meaning of consciousness that the his-torian Yuval Noah Harari had in mind when he wrote, "For every dollar and every minute we invest in improving artificial intelligence, it would be wise to invest a dollar and a minute in advancing human consciousness." The future of self-awareness may mean we end up not only knowing thyself, but also knowing thy machine.[21]

11

EMULATING SOCRATES

> I am not yet able, as the Delphic inscription has it, to know myself; so it seems to me ridiculous, when I do not yet know that, to investigate irrelevant things.
>
> —**PLATO**, *Phaedrus*

We have come full circle. We started our journey through the science of self-awareness with the ancient Athenian call to know ourselves. We now know that this is not an empty platitude. Instead, the human brain is set up to do exactly that. We track uncertainty, monitor our actions, and continually update a model of our minds at work—allowing us to know when our memory or vision might be failing or to encode knowledge about our skills, abilities, and personalities. By understanding how self-awareness works, we have seen how to use it better in situations ranging from the boardroom to the courtroom and begin to extract insights into the machinery and computations that made the human brain self-aware—insights that might inform how we build and interact with AI devices.

In some ways, though, we have ducked Socrates's challenge to discover ways to know ourselves better. It is still early days for research focused on enhancing or improving self-awareness, but there have been some pioneering attempts to develop both high- and low-tech solutions to this challenge. The good news is that, as we saw in Part I of this book, metacognition is not cast in stone but can be shaped and molded by training and experience. This was not always believed to be the case. For much of the twentieth century, the received wisdom among both scientists and the public was that once the human brain had grown up, its circuitry became relatively fixed.

We now know that experience and extensive practice can alter our brain structure well into adulthood. London taxi drivers famously have a larger posterior hippocampus, a region presumably important for storing and retrieving "The Knowledge," the body of information that they are required to learn about London's streets. Skilled musicians show greater gray matter volume in their auditory cortex—again, presumably due to the specialization of this cortical region for processing sounds. There is also evidence that practice can directly lead to changes in brain structure. Learning to juggle leads to increased gray matter volume and white matter in regions of the parietal cortex involved in processing visual motion. In studies of mice, these kinds of volume changes have been linked to the expression of proteins associated with axonal growth, suggesting that brain-volume increases are the observable consequence of extra connections being formed in the brain as a result of practice and training. Extensive research on animals tells us that new learning is accompanied by changes in synaptic weights: the fine-grain molecular structure that allows one neuron to communicate with another.[1]

In other words, not only can adults learn new tricks, but we do not usually have any choice in the matter. The architecture of our brains is being subtly buffeted and updated by everything we do, whether we like it or not. The implication is that, just like other brain functions, metacognition is not fixed or immutable.

To get a sense of what might be possible here, we can take a closer look at experiments that have attempted to modulate or boost metacognition in different ways. In one experiment involving elderly volunteers, researchers at Trinity College Dublin explored whether passing a weak electrical current through the PFC—a technique known as transcranial direct current stimulation (tDCS)—could heighten metacognitive thinking. Subjects were asked to perform a difficult color-detection task under time pressure and to provide a judgment of when they thought they had made an error. In comparison to a sham stimulation condition in which no current was applied, the tDCS increased the volunteers' awareness of when they had made an error, without altering overall performance on the task. We still have little idea of how tDCS works, but it's possible that the weak currents temporarily excite neurons and place the PFC in a state that improves metacognition.[2]

In studies using a similar task, the drug methylphenidate (Ritalin), which increases the concentration of dopamine and noradrenaline, is also able to enhance people's awareness of their errors. My colleagues at UCL have taken this a step further by showing that giving beta-blockers—typically prescribed to reduce blood pressure by blocking noradrenaline function—provides a significant boost in metacognitive sensitivity on a perceptual discrimination task. This was not the case for another group of subjects who instead took a drug designed to block dopamine function, amisulpride. Data on the pharmacology of metacognition is rare, but so far it suggests that boosting systemic levels of dopamine and inhibiting noradrenaline (which, incidentally, is also released in times of stress) may have overall benefits for self-awareness.[3]

It may even be possible to train people to directly alter brain circuits that track confidence in their decisions. In a study published in 2016, researchers at Japan's Advanced Telecommunications Research Institute demonstrated they could train a machine learning algorithm to classify people's confidence levels from the activity of the PFC while they performed a simple task. The researchers then asked people to practice activating those same patterns on their own. If they were able to activate the patterns, a circle on the computer screen got bigger; if they inadvertently began deactivating the patterns, the circle became smaller. There was no need for the subjects to know what they were doing—all they had to do was make the circle bigger, and the algorithm took care of the rest. After two days of training, the participants' metacognition was measured again. In the group that was trained to boost "high confidence" brain patterns, their confidence was increased, whereas in subjects trained to boost "low confidence" brain patterns, confidence was decreased. These represent small changes in metacognitive bias, and because they experienced bigger confidence boosts on incorrect trials, the high-confidence group members in fact suffered in their metacognitive sensitivity. But these findings are a proof of concept that more targeted boosts to self-awareness may be possible.[4]

Many of us might be reluctant to receive brain stimulation or take drugs to boost metacognition. But we might be willing to invest some time in simply practicing being more self-aware. With this goal in mind, my lab has been working on developing training protocols that provide feedback not only on people's performance, but also on their metacognition. We asked people to practice simple perceptual judgments (about which of two images was slightly brighter) for around twenty minutes per day. In one group of subjects, we gave people feedback about their metacognition—whether their judgments of confidence were accurate or inaccurate. A control group of subjects received feedback about their perceptual judgments—whether they chose the correct image. We found that people in the metacognitive feedback group had heightened metacognitive sensitivity after two weeks of regular training.[5]

This promise of being able to improve and reshape our self-awareness depends on whether this kind of training is likely to have benefits that go beyond the laboratory. In our training experiments, improvements in metacognition were also seen in a memory task that was not part of the training. In other words, having learned to boost metacognition on one task, people were naturally able to transfer their newfound skills to better reflect on their performance on a different task. This suggests that the skills we are training may be broad and domain-general. Consistent with a role of metacognition in promoting more open-minded and considered decision-making, we've also found that these self-awareness boosts lead to a systematic improvement in recognizing when you might be wrong and when more information might be useful. There is also some evidence that these boosts in metacognition, installed by practicing a simple game, might be sufficient to promote more open-minded decision-making about contested issues such as climate change.[6]

It is worth sounding a note of caution here. The improvements in metacognition we see in the lab are small, and—aside from potentially beneficial effects on decision-making—are unlikely to have a noticeable impact in our daily lives. However, we should be mindful that more substantial boosts in self-awareness may not be an unalloyed good. In cases such as dementia, a well-intentioned attempt to boost insight and provide patients with awareness of potential memory failures might create unwanted anxiety and depression. Even in the otherwise healthy brain, we have seen that creating more realistic self-appraisals may have emotional downsides, and small doses of overconfidence may be useful for cultivating self-efficacy. Just as we would want to carefully evaluate the side effects of any new drug, we should also be mindful of the potential side effects of boosting self-awareness.[7]

But I hope that by now, I have convinced you that the benefits of improving self-awareness can greatly outweigh the negatives, especially when it is finely tuned. We have seen that self-awareness provides a spark for a whole gamut of human culture—from cave art to philosophy to literary novels—and helps us live in harmony with others, especially when they hold views that oppose our own. Set against this background, Socrates's suggestion that we should spend time boosting our self-awareness seems more relevant than ever.

More Than Conscious

If it were possible to systematically improve or alter self-awareness, what would that feel like? The answer to that question depends on how we think about the relationship between metacognition and awareness of the world around us. To see this, imagine that I put you in a dark room and asked you to watch for a faint light to appear on a computer screen. When the light appears, it presumably causes a difference in your visual experience. But it also leads to a difference in self-awareness: you think about whether you saw the stimulus or not, and this reflective judgment allows you to notice and communicate a change in consciousness. How should we think about this relationship?

There are two dominant camps in the field. The first approach is to suggest that self-awareness is just an optional veneer on regular conscious experience. This "first-order" view accepts that we might need metacognition to reflect on and report our experience, but argues that this does not mean it is involved in creating the experience itself. Instead, metacognition is an optional add-on, layered on top of consciousness and merely required for getting the information out into the outside world.

One argument in support of the first-order view is known as the overflow thesis, which goes something like this: The world around you appears rich and detailed. From my desk, I can pick out the changing shades of white on the wallpaper as the light from my Anglepoise lamp plays off the wall, the wood pattern of my desk, and the deep blue of a painting on the wall. I could go on, but it seems obvious that there is far more in this scene than I can talk about. My private inner world is poorly served by our ability to access and describe it—it "overflows" the capacity for report.[8]

Overflow seems intuitive, and it has been used to support the idea that our experience and our ability to comment on that experience are two different things. But we should be careful before accepting this conclusion. Data from overflow experiments is also consistent with us having only a vague impression of a whole set of objects, which is brought into clearer focus by the requirement for report. It's fiendishly difficult to know for sure what people's primary conscious experience is like without asking them to reflect on what they are experiencing.[9]

An alternative camp is known as the "higher-order" view. Higher-order theorists propose that being able to reflect on and generate thoughts about our mental states is exactly what makes us conscious in the first place. Without metacognitive awareness of our mental states, we might still be able to process information or react to stimuli, but there would be no consciousness to speak of.[10]

There are some key predictions that higher-order and first-order theorists disagree on. For instance, to the extent that the PFC is important for metacognition and higher-order thought, then if two experimental conditions are created that differ only in conscious awareness, you should detect a difference in the PFC under the higher-order view but not under the first-order view. There is initial data that supports this hypothesis. The problem, though, is that the technologies we currently have for studying the human brain (such as fMRI and MEG) are too coarse to provide a strong pronouncement either way. We do not yet have unfettered access to the fine-grain patterns of neural activity underpinning subtle changes in the conscious and metacognitive states of human subjects—and so the jury remains out.[11]

There is, however, at least circumstantial evidence for a connection between metacognition and consciousness. The studies of blindsight we encountered earlier suggest a lack of metacognitive sensitivity is a feature of information processing in the unconscious (blind) hemifield. Conversely, damage to the PFC does not only impair metacognition, but it can also have consequences for conscious experience.[12]

Another, more subtle implication is that if metacognition is inherent to conscious experience, it may be difficult for us to reflect on consciousness itself. We have already seen that we rely on metacognition to distinguish between reality and imagination, to realize that our perceptual experience may be in error, and to recognize that we might be fooled by illusions. We may look at a visual illusion such as the checkerboard we encountered in Part I, and say to ourselves, "I *know* the squares are the same shade of gray, but it still seems like they are different." I can recognize that what I see and what I think I see are in conflict. Encoding these differences between perception and reality allows us to have an awareness of seeing—an important component of subjective experience. But it seems much more difficult to make consciousness itself visible in this way. To us, consciousness just is—it is transparent. This transparency may even be the root of the hard problem of consciousness: why we think that there is a mystery of consciousness at all.[13]

My personal view is that this debate will hinge on which version of consciousness we care about more. It certainly seems possible that primary consciousness could exist in a vacuum, without our ability to introspect about it—to have a raw experience of smelling coffee, or seeing the color red, without engaging in metacognition at all. However, I also suspect that the kind of consciousness we cherish—the kind that allows us to appreciate the smell of coffee and tell our friends about the sunset—involves meta-awareness, that is, a state of knowing that we are conscious. This higher level of awareness is not just an optional veneer or gloss on experience, but instead a base layer of what it means to be human.[14]

Consciously experiencing emotions may also involve this kind of higher-order awareness. The neuroscientist Joseph Le-Doux suggests that the bodily reactions that often accompany emotional states—such as freezing in response to a loud sound or sweating in response to a snake or spider—may be distinct to the conscious, reflective experience of shock or fear. He points out that laboratory mice and humans show similar automatic reactions governed by circuits deep in the brain when faced with danger, but it is likely that only humans can think about and become aware of what it means to feel fear.[15]

An interesting test for exploring the meaning of these different varieties of conscious experience is the puzzling case of dreams. It certainly seems that we are conscious in our dreams. But at the time of dreaming, it is rare that we have any self-awareness of being in a dream. This renders the experience a bit frustrating and meaningless, at least until we wake up. Wouldn't it be great if we could sometimes become conscious of our dreams?

In fact, some people do report experiencing self-awareness during dreaming, in states known as lucid dreams. Lucid dreams tend to occur when people are overly tired, though it may also be possible to train yourself to become lucid at will. The frequency of lucid dreams also varies substantially between people—some never experience them, others experience around one lucid dream per month, and yet others dream lucidly most nights. I have only experienced a lucid dream once, when I was sleeping on a boat moored in the Hamble River on the south coast after a few days' sailing. I suddenly became aware of flying through the boat and being able to guide myself around at will. It was a bizarre and wonderful experience. It felt like being fully awake and conscious, but in another world.

It is possible to track whether someone has gone into a lucid state by asking them to make a telltale set of eye movements that can be monitored while the person is asleep. Remarkably, brain imaging studies have shown that activity in the frontopolar cortex and precuneus is increased during lucid dreams—two regions that we have seen are also implicated in metacognition. Even more remarkably, electrical stimulation of the PFC at a particular frequency is sufficient to increase people's lucidity during their dreams.[16]

Research on lucid dreaming, then, is perfectly consistent with the notion that people are not meaningfully aware of their dreams under normal circumstances. As in the case of overflow, it remains possible that we have something like primary consciousness without self-awareness during dreams, but it would seem difficult or impossible to tell whether this is the case with the current measures we have available. Instead, the data shows us that when people become conscious of their dream—when they become lucid—they recruit the same brain networks that support waking metacognition.

Becoming Lucid

I find it appealing that boosts to self-awareness in our daily lives might feel like the experience of becoming lucid in a dream—we might begin to notice things that we have not noticed before in ourselves, in others, and in the world. This is in fact strikingly close to the kinds of changes in consciousness that expert meditators report experiencing after intensive retreats.

It might be no surprise, then, that another promising route toward boosting metacognition and self-awareness is by engaging in regular meditation. Mindfulness meditation is central to Buddhist practice and has been closely linked to the kind of higher-order, reflective awareness that is characteristic of meta-cognition. But the impact of meditation on scientific metrics of self-awareness has only recently been explored. The data is encouraging. A 2014 study led by psychologists Benjamin Baird and Jonathan Schooler at University of California, Santa Barbara, reported that engaging in two weeks of meditation training increased metacognitive sensitivity during a memory test. Other work has shown that meditation experts have greater metacognitive sensitivity compared to controls.[17]

This is a small but growing area of research, and much remains to be done to replicate and extend these findings. Meditation studies have proven controversial in neuroscience because it is often difficult to create agreed-upon definitions of what meditation training means. But I remain optimistic. One exciting idea is that because mindfulness meditation involves consistent self-focus and the ability to zero in on our mental states, it might also hone our ability for self-appraisal. And the power of periods of reflection for boosting performance need not be restricted to classical meditation. Researchers at Harvard Business School compared groups of trainee employees at the Indian IT company Wipro, assigning them to spend the last fifteen minutes of their day either reflecting on what they had learned (the reflection condition), explaining the main lessons to others (the sharing condition), or continuing their studies as normal (the control condition). Compared to the control condition, the employees in both the reflection and sharing conditions boosted their performance on a final exam by over 20 percent.[18]

We have already seen, though, that self-awareness and meta-cognition are fragile, and the scope of plasticity is asymmetric. In other words, the gains that we can achieve by boosting meta-cognition are nothing compared to the downsides of losing it. It is concerning, then, that self-awareness may be under increasing threat from modern life. In a culture of being efficient and productive, taking time to reflect on what we are doing seems to be an optional luxury. Smartphones and screens are dominating our waking moments and squeezing out times at which we otherwise might have stopped to think. As we have seen, excess stress and deterioration in mental health may also lead to the erosion of self-awareness. There is a danger that we will become trapped in a vicious cycle in which we are more focused on doing and less on reflecting, and become less and less aware of the benefits of high-quality metacognition. By understanding the factors that lead to self-awareness failure, we can take steps to stop this cycle before it begins.

In turn, a science of self-awareness helps us adopt a compassionate stance toward occasional failures of self-awareness in others. Just as for ourselves, the self-awareness of our friends and colleagues is continually in flux and buffeted by a range of hidden signals affecting whether they think they are right or wrong about a particular issue. People who hold opposite views to our own on polarizing issues ranging from politics to vaccines may appear "blind" to the evidence. But by recognizing that our confidence in our views is a construction, and prone to distortion, we can cultivate a more tolerant attitude toward others who may not agree with us.

Perhaps the most important element in protecting and cultivating self-awareness is something you are just about to finish doing: reading and thinking about the science of self-awareness. By pulling back the veil, even for a moment, on how metacognition works, we can gain a newfound respect for the fragility and power of the reflective mind. There is a beautiful symmetry to this. Simply by studying self-awareness, we may gain more of it. Two and a half thousand years after the Athenians wrote their words of advice on the Temple of Delphi, we are now in a better position than ever before to know ourselves.

ACKNOWLEDGMENTS

One of the wonderful things about doing science is that it involves a constant exchange of ideas with friends and colleagues. This means it's also a nearly impossible task to properly acknowledge everyone who has contributed to the ideas in this book and the many colleagues who have created the field of metacognitive neuroscience. I am especially indebted to Paul Azzopardi, who, during an exhilarating eight-week undergraduate course, first showed me that a science of subjective experience was not only possible but well underway in the form of research on blindsight. I've been lucky to have superb PhD mentorship from Chris Frith and Ray Dolan, who provided both the freedom to explore and gentle but important words of advice that nudged our work on metacognition toward the most interesting questions. The vibrant international community of metacognition and consciousness researchers has become like a second family, with the meetings of the Association of the Scientific Study of Consciousness as our annual gatherings.

The Wellcome Trust, Royal Society, and Leverhulme Trust have not only provided generous support for much of the research I discuss in the book, but also fostered a culture in the UK in which public engagement and science communication are championed and considered integral to the enterprise of science. It's been a privilege to be surrounded by kind yet incisive colleagues at University College London, and to be part of a dynamic network of cognitive neuroscientists and psychologists at the Wellcome Centre for Human Neuroimaging, the Max Planck Centre for Computational Psychiatry and Ageing Research, and the Department of Experimental Psychology. Likewise, members of my lab, both past and present, have been a wellspring of ideas and enthusiasm. My thanks go to PhD students Max Rollwage, Matan Mazor, Elisa van der Plas, Xiao Hu, and Andrew McWilliams; postdoctoral colleagues Marion Rouault, Dan Bang, Jon Huntley, Nadim Atiya, and Nadine Dijkstra; and research assistants or affiliate students Jason Carpenter, Oliver Warrington, Jihye Ryu, Sara Ershadmanesh, Audrey Mazancieux, Anthony Vaccaro, and Tricia Seow. Outside of the lab, Hakwan Lau and Benedetto De Martino have patiently listened to my many half-baked ideas and worries, and they have been both sounding boards and true friends in science. I have also been lucky to build a partnership with Dan Schaffer, whose stimulating discussion evenings with his legal trainees at Slaughter and May have sharpened my thoughts on the real-world relevance of metacognition.

The writing took place over three consecutive summers, and different chapters are indelibly linked to specific places. Thibault Gajdos kindly hosted me as a visiting researcher at Aix-Marseille Université during the summer of 2018, where I gave a series of lectures that formed the basis of Part I. Much of the remainder was completed in the Scottish village of Crail while on paternity leave the following year, and I'm grateful to my parents-in-law for use of their wonderful flat above the harbor there. The final chapters have been written in Zagreb, Croatia, at the start of my wife's diplomatic posting. Throughout these moves, the ever-present support of our parents and friends has been central to keeping us emotionally, if not always physically, tied to home.

Writing this book has been an acutely metacognitive experience—accompanied by self-doubt, self-questioning, and second-guessing of whether things are on the right or wrong track. The onset of a global pandemic in the closing stages only heightened this introspective anxiety, and I am very grateful to those at the other end of an email for prompt advice and words of encouragement. I am particularly indebted to Chris Frith and Nicholas Shea, who read the final draft and provided incisive and timely comments that helped minimize some of my more egregious errors and omissions. My editors at *Scientific American* and *Aeon* magazines, Sandra Upson and Brigid Hains, were instrumental in shaping my initial ideas on metacognition into a coherent narrative. My agent, Nathaniel Jacks, has been incredibly patient and helpful, particularly during those New York days during which the ideas for this book developed in fits and starts. My editors at Basic Books and John Murray, TJ Kelleher and Georgina Laycock, have combined timely nudges on structure and content with calming cups of tea during our London get-togethers. The sharp eye and error awareness of Liz Dana proved invaluable in the latter stages. For reading and commenting on early drafts and individual chapters, I'm grateful to Oliver Hulme, Nicholas Wright, Benedetto De Martino, Cecilia Heyes, Dan Bang, Katerina Fotopoulou, Robert Rothkopf, Will Robinson, Alex Fleming, and Helen Walker-Fleming.

My wife, Helen, has provided unstinting love, support, and patience, and my gratitude knows no bounds. Finn, watching you grow up and acquire the very thing I was trying to research and write about has been awe-inspiring and humbling in equal measure. This book is for both of you.

NOTES

Preface

1. Linnaeus (1735); Flavell (1979); Nelson and Narens (1990); Metcalfe and Shimamaura (1996).

2. Nestor (2014).

3. Shimamura (2000); Fleming and Frith (2014).

4. The MetaLab, https://metacoglab.org.

5. Comte (1988).

6. Descartes (1998).

7. Mill (1865).

8. Dennett (1996).

9. From a BBC interview with James Mossman, published in Vladimir Nabokov, *Strong Opinions* (New York: Vintage, 1990).

10. Hamilton, Cairns, and Cooper (1961).

11. Renz (2017).

12. Baggini (2019); Ivanhoe (2003).

13. Dennett (2018).

14. The terminology used by scientists and philosophers studying self-awareness and metacognition can get confusing at times. In this book, I use the terms *metacognition* and *self-monitoring* to refer to any process that monitors another cognitive process, such as realizing we have made an error in solving a math problem. Self-monitoring and metacognition may sometimes occur unconsciously. I reserve the term *self-awareness* for the ability to consciously reflect on ourselves, our behavior, and our mental lives. Some psychologists restrict the term self-awareness to mean bodily self-awareness, or awareness of the location and appearance of the body, but here I am generally concerned with awareness of mental states.

Chapter 1: How to Be Uncertain

1. Jonathan Steele, "Stanislav Petrov Obituary," *The Guardian*, October 11, 2017, www.theguardian.com/world/2017/oct/11/stanislav-petrov-obituary.

2. Green and Swets (1966).

3. The seeds of Bayes's rule were first identified by the eleventh-century Arabic mathematician Ibn al-Haytham, developed by English clergyman and mathematician Thomas Bayes in 1763, and applied to a range of scientific problems by the eighteenth-century French mathematician Pierre-Simon Laplace. See McGrayne (2012).

4. Felleman and Van Essen (1991); Zeki and Bartels (1998).

5. Clark (2013); Clark (2016); Craik (1963); Friston (2010); Helmholtz (1856); Gregory (1970); Hohwy (2013).

6. Kersten, Mamassian, and Yuille (2004); Ernst and Banks (2002); Pick, Warren, and Hay (1969); Bertelson (1999); McGurk and MacDonald (1976).

7. Born and Bradley (2005); Ma et al. (2006).

8. Apps and Tsakiris (2014); Blanke, Slater, and Serino (2015); Botvinick and Cohen (1998); Della Gatta et al. (2016); Seth (2013).

9. Kiani and Shadlen (2009); Carruthers (2008); Insabato, Pannunzi, and Deco (2016); Meyniel, Sigman, and Mainen (2015).

10. Smith et al. (1995).

11. It's possible that alternative explanations that do not require tracking uncertainty could account for animals' behavior in these experiments. For instance, when the third lever is introduced to Natua, there are now three responses: low tone, high tone, and "don't know" (the opt-out response). After a while, Natua might learn that pressing the low or high keys when the tone is in the middle will often lead to a penalty for getting the wrong answer, and no fish to eat. The opt-out response is less risky, allowing him to move on swiftly to another trial where fish are on the table again. He might just be following a simple rule, something like, "When the middle tone occurs, press the opt-out lever," without feeling uncertain about whether he's likely to get the answer right. Carruthers (2008).

12. Kornell, Son, and Terrace (2007); Shields et al. (1997); Kepecs et al. (2008); Fujita et al. (2012). Six pigeons and two out of three bantam chickens were more likely to use the risky option when they were correct on a visual search task. Two pigeons also showed consistent generalization of this metacognitive ability to new sets of colors.

13. Goupil and Kouider (2016); Goupil, Romand-Monnier, and Kouider (2016).

14. The psychologist Josep Call offers the following summary: "I think that it is perhaps fair to say that the field has entered a sort of arms race in which increasingly elaborated non-metacognitive explanations are met with ever more sophisticated empirical evidence which in turn generate increasingly more complex non-metacognitive explanations." Call (2012); Hampton (2001).

15. Beran et al. (2009).

16. Meyniel, Schlunegger, and Dehaene (2015).

17. The Hungarian mathematician Abraham Wald developed the theory of sequential analysis while working for the US government during World War II. Turing independently developed similar methods as part of the Banburismus process, which remained classified by the UK government until the 1980s. Hodges (1992); Wald (1945); Gold and Shadlen (2002).

18. Desender, Boldt, and Yeung (2018); Desender et al. (2019).

19. Bayesian inference is straightforward in restricted situations with only a few hypotheses. But when the problem becomes unconstrained, there is an explosion of possible dimensions along which we must estimate probabilities, making it rapidly intractable. A fast-moving research field spanning both AI and cognitive science is working on ever more ingenious approximations to Bayesian inference, and it is possible that similar approximations are used by the brain.

Chapter 2: Algorithms for Self-Monitoring

1. Allostasis refers to the process of predicting how homeostasis will need to be adjusted: "Stability through change." Conant and Ashby (1970); Sterling (2012).

2. Clark (2016); Hohwy (2013); Pezzulo, Rigoli, and Friston (2015); Gershman and Daw (2012); Yon, Heyes, and Press (2020).

3. Badre and Nee (2018); Passingham and Lau (2019).

4. Michael Church, "Method & Madness: The Oddities of the Virtuosi," *The Independent*, March 12, 2008, www.independent.co.uk/arts-entertainment /music/features/method-madness-the-oddities-of-the-virtuosi-794373.html.

5. Logan and Zbrodoff (1998); Logan and Crump (2011).

6. Crump and Logan (2010); Logan and Crump (2009).

7. Reynolds and Bronstein (2003).

8. Fourneret and Jeannerod (1998).

9. Diedrichsen et al. (2005); Schlerf et al. (2012). There is an alternative view in which this copy is not secondary to the main command—it (or at least part of it) *is* the command. This is known as active inference. At the heart of active inference is a deep symmetry between perceptual and motor prediction errors. Perceptual prediction errors alter our model of the world; in turn, motor or "proprioceptive" prediction errors cause our muscles to move to shape our limbs to be in line with our predictions. In other words, we induce an error by saying, "I want to (expect to) be over there," and our motor system snaps into line. Clark (2013); Friston (2010); Adams, Shipp, and Friston (2013); Friston et al. (2010).

10. Blakemore, Wolpert, and Frith (2000); Shergill et al. (2003); Wolpert and Miall (1996).

11. Rabbitt (1966); Rabbitt and Rodgers (1977); Hasbroucq et al. (1999); Meckler et al. (2017).

12. Gehring et al. (1993); Dehaene, Posner, and Tucker (1994); Fu et al. (2019).

13. Goupil and Kouider (2016).

14. Schultz, Dayan, and Montague (1997). Associative learning comes in different forms. In "classical" or Pavlovian conditioning, anticipatory responses come to be associated with a stimulus or cue. In "operant" or instrumental conditioning, the animal or human needs to perform an action in order to get a reward.

15. Seymour et al. (2004); O'Doherty et al. (2003); Sutton and Barto (2018). Prediction errors are a key mathematical variable needed to train learning algorithms in a branch of computer science known as reinforcement learning (RL). RL suggests that when learning is complete, no additional dopamine needs to be released, just as Schultz found: the monkey has come to expect the juice after the light, and there is no more error in his prediction. But it also predicts that if the juice is unexpectedly taken away, the baseline dopamine response dips—a so-called negative prediction error. This was also borne out by the neuronal recordings.

16. Another way of thinking about the role of the dACC and signals like the ERN is that they signal intermediate progress on the way toward obtaining a more concrete or explicit reward. Botvinick, Niv, and Barto (2009); Shidara and Richmond (2002); Ribas-Fernandes et al. (2011).

17. Gadagkar et al. (2016); Hisey, Kearney, and Mooney (2018).

18. There is unlikely to be a sharp division between these different levels. For instance, the ERN is itself modulated by the smoothness of the action we make to get to the target. Torrecillos et al. (2014).

19. See Stephen M. Fleming, "False Functional Inference: What Does It Mean to Understand the Brain?" *Elusive Self* (blog), May 29, 2016, https://elusiveself.wordpress.com/2016/05/29/false-functional-inference-what-does-it-mean-to-understand-the-brain/; and Jonas and Kording (2017); Marr and Poggio (1976).

20. Another note on terminology is useful here. The philosopher Joëlle Proust distinguishes between procedural and analytic metacognition: procedural metacognition is based on lower-level feelings of fluency that may or may not be conscious, whereas analytic metacognition is based on reasoning about one's own competences. Others, for instance Peter Carruthers, deny that implicit monitoring and control qualify as metacognition because they can be explained without appeal to meta-representation. Still others, such as Josef Perner, accept the primacy of meta-representation as a starting point for thinking about metacognition, but are willing to allow a gradation of mental processes that are intermediate in level between implicit monitoring and full-blown, conscious meta-representation. Perner (2012); Proust (2013); Carruthers (2008).

Chapter 3: Knowing Me, Knowing You

1. Aubert et al. (2014); McBrearty and Brooks (2000); Sterelny (2011).

2. Ryle (2012).

3. Carruthers (2009); Carruthers (2011); Fleming and Daw (2017); Thornton et al. (2019).

4. Baron-Cohen, Leslie, and Frith (1985); Wimmer and Perner (1983).

5. Hembacher and Ghetti (2014).

6. Bretherton and Beeghly (1982); Gopnik and Astington (1988); Flavell (1979); Rohwer, Kloo, and Perner (2012); Kloo, Rohwer, and Perner (2017); Filevich et al. (2020).

7. Lockl and Schneider (2007); Nicholson et al. (2019); Nicholson et al. (2020).

8. Darwin (1872); Lewis and Ramsay (2004); Kulke and Rakoczy (2017); Onishi and Baillargeon (2005); Scott and Baillargeon (2017); Paulus, Proust, and Sodian (2013); Wiesmann et al. (2020).

9. Courage, Edison, and Howe (2004). An alternative perspective is that the mirror test is probing a distinct (and possibly nonconscious) ability to use mirrors appropriately (as we do effortlessly when shaving or doing our hair) without requiring self-awareness. Heyes (1994); Chang et al. (2017); Kohda et al. (2019).

10. Bretherton and Beeghly (1982); Gopnik and Astington (1988).

11. Lewis and Ramsay (2004).

12. Call and Tomasello (2008); Kaminski, Call, and Tomasello (2008); Butterfill and Apperly (2013); Heyes (2015); Krupenye and Call (2019); Premack and Woodruff (1978).

13. Herculano-Houzel, Mota, and Lent (2006); Herculano-Houzel et al. (2007).

14. Birds also appear to have a primate-like scaling law. Dawkins and Wong (2016); Herculano-Houzel (2016); Olkowicz et al. (2016).

15. This raises an obvious question. We might have big heads, but we certainly don't have the biggest *brains* around. What about huge but evolutionarily distant species, such as elephants or whales? Herculano-Houzel found that, in fact, an African elephant brain is not just larger than the human brain, it also has three times the number of neurons. At first glance, this would appear to be a spanner in the works of the theory that humans have unusually large numbers of neurons compared to other species. But it turns out that the vast majority—98 percent—of the elephant's neurons are located in its cerebellum, not the cortex. As we saw in the previous chapter, it is likely that the cerebellum acts as a suite of autopilots, keeping action and thought on track, but (in humans at least) not generating any kind of self-awareness. It is possible that the elephant needs such a large cerebellum due to its complex body plan and trunk that requires a lot of fine motor management. The African elephant is the exception that proves the rule of human uniqueness;

humans retain a cortical neuronal advantage relative to any species tested to date. Herculano-Houzel et al. (2014).

16. Ramnani and Owen (2004); Mansouri et al. (2017); Wallis (2011); Semendeferi et al. (2010).

17. Jenkins, Macrae, and Mitchell (2008); Mitchell, Macrae, and Banaji (2006); Ochsner et al. (2004); Kelley et al. (2002); Northoff et al. (2006); Lou, Changeux, and Rosenstand (2017); Summerfield, Hassabis, and Maguire (2009).

18. Lou et al. (2004).

19. Shimamura and Squire (1986).

20. Janowsky et al. (1989); Schnyer et al. (2004); Pannu and Kaszniak (2005); Fleming et al. (2014); Vilkki, Servo, and Surma-aho (1998); Vilkki, Surma-aho, and Servo (1999); Schmitz et al. (2006); Howard et al. (2010); Modirrousta and Fellows (2008).

21. Nelson et al. (1990).

22. Kao, Davis, and Gabrieli (2005).

23. Amodio and Frith (2006); Vaccaro and Fleming (2018).

24. Armin Lak and Adam Kepecs have shown that neuronal firing in the rat frontal cortex predicts how long they are willing to wait for a reward for getting the answer right on a decision-making task—a marker of implicit metacognition. Inactivating this same region by infusing a drug known as muscimol impairs their ability to wait, but not to make the initial decision. In this aspect, the rodents in Lak and Kepecs's study are similar to humans with damage to the prefrontal cortex: their cognition was intact, but their metacognition was impaired. Other work in monkeys has shown that neurons in the parietal and frontal lobes and thalamus track uncertainty about the evidence in support or different decisions, such as whether a stimulus is moving to the left or right—just as Turing's equations tracked evidence for or against a particular Enigma hypothesis. Lak et al. (2014); Middlebrooks and Sommer (2012); Kiani and Shadlen (2009); Miyamoto et al. (2017); Miyamoto et al. (2018); Komura et al. (2013).

25. Mesulam (1998); Krubitzer (2007).

26. Margulies et al. (2016); Baird et al. (2013); Christoff et al. (2009); Passingham, Bengtsson, and Lau (2010); Metcalfe and Son (2012); Tulving (1985).

27. Herculano-Houzel (2016).

Chapter 4: Billions of Self-Aware Brains

1. Freud (1997); Mandler (2011).

2. Like many potted histories, this one is oversimplified. It is clear that the first psychologists studying subjective aspects of the mind also took behavior seriously, and Wundt himself ended up being one of the harshest critics of

research on introspection conducted by his students Titchener and Külpe. Conversely, research on animal behavior was well underway before the advent of behaviorism (Costall, 2006). There were also some early glimmers of modern approaches to quantifying the accuracy of self-awareness: in a classic paper that was ahead of its time, the Victorian psychologists Peirce and Jastrow proposed a mathematical model of metacognition, suggesting that $m = c\log \frac{p}{1-p}$ where m denoted the confidence level of the subject, p is the probability of an answer being right, and c is a constant (Peirce and Jastrow, 1885). This equation stated that subjects' confidence goes up in proportion to the logarithm of the probability of being right—an assertion supported by recent experiments. Van den Berg, Yoo, and Ma (2017).

3. Hart (1965).

4. There comes a point at which bias and sensitivity collide—if I am always 100 percent confident, then I will tend to have a high bias *and* low sensitivity. Clarke, Birdsall, and Tanner (1959); Galvin et al. (2003); Nelson (1984); Maniscalco and Lau (2012); Fleming and Lau (2014); Fleming (2017); Shekhar and Rahnev (2021).

5. Fleming et al. (2010).

6. Poldrack et al. (2017).

7. Yokoyama et al. (2010); McCurdy et al. (2013). See also Fleming et al. (2014); Hilgenstock, Weiss, and Witte (2014); Miyamoto et al. (2018); Baird et al. (2013); Baird et al. (2015); Barttfeld et al. (2013); Allen et al. (2017); Rounis et al. (2010); Shekhar and Rahnev (2018); Qiu et al. (2018).

8. Semendeferi et al. (2010); Neubert et al. (2014).

9. Cross (1977); Alicke et al. (1995). In a phenomenon known as the Dunning-Kruger effect, after its discoverers, overconfidence biases are most pronounced in those who perform poorly (Dunning, 2012; Kruger and Dunning, 1999). Kruger and Dunning propose that low performers suffer from a metacognitive error and not a bias in responding (Ehrlinger et al., 2008). However, it is still not clear whether the Dunning-Kruger effect is due to a difference in metacognitive sensitivity, bias, or a mixture of both. See Tal Yarkoni, "What the Dunning-Kruger Effect Is and Isn't," *[citation needed]* (blog), July 7, 2010, https://talyarkoni.org/blog/2010/07/07/what-the-dunning-kruger-effect-is-and-isnt; and Simons (2013).

10. Ais et al. (2016); Song et al. (2011).

11. Mirels, Greblo, and Dean (2002); Rouault, Seow, Gillan, and Fleming (2018); Hoven et al. (2019).

12. Fleming et al. (2014); Rouault, Seow, Gillan, and Fleming (2018); Woolgar, Parr, and Cusack (2010); Roca et al. (2011); Toplak, West, and Stanovich (2011); but see Lemaitre et al. (2018).

13. Fleming et al. (2015); Siedlecka, Paulewicz, and Wierzchoń (2016); Pereira et al. (2020); Gajdos et al. (2019).

14. Logan and Crump (2010).

15. Charles, King, and Dehaene (2014); Nieuwenhuis et al. (2001); Ullsperger et al. (2010).

16. Allen et al. (2016); Jönsson, Olsson, and Olsson (2005).

17. De Gardelle and Mamassian (2014); Faivre et al. (2018); Mazancieux et al. (2020); Morales, Lau, and Fleming (2018). There are also intriguing exceptions to domain-generality that deserve further study. First of all, just because metacognition appears to be domain-general from the vantage point of behavior does not mean that different metacognitive abilities depend on the same neural circuitry (see, for example, McCurdy et al., 2013; Baird et al., 2013; Baird et al., 2015; Fleming et al., 2014; Ye et al., 2018). Second, some modalities do appear to be metacognitively special—Brianna Beck, Valentina Peña-Vivas, Patrick Haggard, and I have found that variability in metacognition about painful stimuli does not predict metacognition for touch or vision, despite the latter two abilities being positively correlated. Beck et al. (2019).

18. Bengtsson, Dolan, and Passingham (2010); Bandura (1977); Stephan et al. (2016); Rouault, McWilliams, Allen, and Fleming (2018); Will et al. (2017); Rouault, Dayan, and Fleming (2019); Rouault and Fleming (2020).

19. Bang and Fleming (2018); Bang et al. (2020); Fleming and Dolan (2012); Fleming, Huijgen, and Dolan (2012); Gherman and Philiastides (2018); Passingham, Bengtsson, and Lau (2010); De Martino et al. (2013); Fleming, van der Putten, and Daw (2018).

20. Heyes and Frith (2014); Heyes (2018).

21. Pyers and Senghas (2009); Mayer and Träuble (2013).

22. Hughes et al. (2005). Similar studies of the genetics of metacognition are rare. David Cesarini at New York University studied 460 pairs of Swedish twins and asked them to carry out a twenty-minute test of general cognitive ability. Before taking the test, each twin also rated how they believed they would rank relative to others—a measure of over- or underconfidence. The study found that genetic differences explained between 16 percent and 34 percent of the variation in overconfidence (Cesarini et al., 2009). Similar results were obtained in a larger study of more than 7,500 children in the UK by Corina Greven, Robert Plomin, and colleagues at Kings College London. They collected both confidence data—How good do the children think they are at English, science, and mathematics?—and also measures of IQ and actual success at school. Strikingly, the results showed that around half the variance in children's confidence level was influenced by genetic factors—around the same degree as the genetic influence on IQ itself (Greven et al., 2009). So far, genetic studies have only examined our overall sense of confidence in our abilities, and none have quantified metacognitive prowess using the tools described in this chapter. It would be fascinating to apply similar techniques to look at variation in metacognitive sensitivity: Does our genetic makeup also predict how well we know our own minds? Or is metacognition more like mindreading, a skill that is less tied to our genes and more influenced by the rich thinking tools we acquire from our parents and teachers?

23. Heyes et al. (2020).

24. Weil et al. (2013).

25. Blakemore (2018); Fandakova et al. (2017).

26. David et al. (2012).

27. Fotopoulou et al. (2009).

28. Marsh (2017).

29. Burgess et al. (1998); Schmitz et al. (2006); Sam Gilbert and Melanie George, "Frontal Lobe Paradox: Where People Have Brain Damage but Don't Know It," *The Conversation*, August 10, 2018, https://the conversation.com/frontal-lobe-paradox-where-people-have-brain-damage -but-dont-know-it-100923.

30. Cosentino (2014); Cosentino et al. (2007); Moulin, Perfect, and Jones (2000); Vannini et al. (2019); Agnew and Morris (1998); Morris and Mograbi (2013).

31. Johnson and Raye (1981); Simons, Garrison, and Johnson (2017).

32. Frith (1992); Knoblich, Stottmeister, and Kircher (2004); Metcalfe et al. (2012).

33. Harvey (1985); Bentall, Baker, and Havers (1991); Garrison et al. (2017); Simons et al. (2010).

34. Eichner and Berna (2016); Moritz and Woodward (2007); Moritz et al. (2014).

Chapter 5: Avoiding Self-Awareness Failure

1. Alter and Oppenheimer (2006); Alter and Oppenheimer (2009); Reber and Schwarz (1999); Hu et al. (2015); Palser, Fotopoulou, and Kilner (2018); Thompson et al. (2013).

2. Kahneman (2012).

3. Schooler et al. (2011).

4. Smallwood and Schooler (2006).

5. Goldberg, Harel, and Malach (2006).

6. Reyes et al. (2015); Reyes et al. (2020).

7. Metzinger (2015).

8. Metcalfe and Finn (2008); Kornell and Metcalfe (2006); Rollwage et al. (2020); Koizumi, Maniscalco, and Lau (2015); Peters et al. (2017); Samaha, Switzky, and Postle (2019); Zylberberg, Barttfeld, and Sigman (2012).

Chapter 6: Learning to Learn

1. "Equipping People to Stay Ahead of Technological Change," *The Economist*, January 14, 2017, www.economist.com/leaders/2017/01/14/equip ping-people-to-stay-ahead-of-technological-change.

2. Christian Jarrett et al., "How to Study and Learn More Effectively," August 29, 2018, in *PsyCruch*, produced by Christian Jarrett, podcast, 13:00, https://digest.bps.org.uk/2018/08/29/episode-13-how-to-study-and-learn -more-effectively/; Christian Jarrett, "All You Need to Know About the

'Learning Styles' Myth, in Two Minutes," *Wired*, January 5, 2015, www.wired
.com/2015/01/need-know-learning-styles-myth-two-minutes/.

3. Knoll et al. (2017).

4. Ackerman and Goldsmith (2011).

5. Bjork, Dunlosky, and Kornell (2013); Kornell (2009); Kornell and Son
(2009); Karpicke (2009); Zimmerman (1990).

6. Dunlosky and Thiede (1998); Metcalfe and Kornell (2003); Metcalfe
and Kornell (2005); Metcalfe (2009).

7. Schellings et al. (2013); De Jager, Jansen, and Reezigt (2005); Jordano
and Touron (2018); Michalsky, Mevarech, and Haibi (2009); Tauber and
Rhodes (2010); Heyes et al. (2020).

8. Chen et al. (2017).

9. Diemand-Yauman, Oppenheimer, and Vaughan (2011); "Sans Forgetica," RMIT University, https://sansforgetica.rmit.edu.au/.

10. In 2016 the penalty for guessing was removed, see *Test Specifications
for the Redesigned SAT* (New York: College Board, 2015), 17–18, https://
collegereadiness.collegeboard.org/pdf/test-specifications-redesigned-sat
-1.pdf. Ironically, this change in the rules may have inadvertently removed
the (also presumably unintended) selection for metacognition inherent to the
previous scoring rule. See Higham (2007).

11. Koriat and Goldsmith (1996).

12. Bocanegra et al. (2019); Fandakova et al. (2017).

13. Bandura (1977); Cervone and Peake (1986); Cervone (1989); Weinberg, Gould, and Jackson (1979); Zacharopoulos et al. (2014).

14. Greven et al. (2009); Chamorro-Premuzic et al. (2010); Programme
for International Student Assessment (2013).

15. Kay and Shipman (2014).

16. Clark and Chalmers (1998); Clark (2010); Risko and Gilbert (2016);
Gilbert (2015); Bulley et al. (2020).

17. Hu, Luo, and Fleming (2019).

18. Ronfard and Corriveau (2016).

19. Csibra and Gergely (2009); Lockhart et al. (2016).

20. Bargh and Schul (1980); Eskreis-Winkler et al. (2019).

21. Trouche et al. (2016); Sperber and Mercier (2017).

22. Koriat and Ackerman (2010).

23. Clark (2010); Mueller and Oppenheimer (2014).

Chapter 7: Decisions About Decisions

1. Mark Lynas, interview with Dominic Lawson, February 4, 2015, in
Why I Changed My Mind, produced by Martin Rosenbaum, podcast, 15:30,
www.bbc.co.uk/sounds/play/b0510gvx.

2. Van der Plas, David, and Fleming (2019); Fleming (2016).

3. Fleming, van der Putten, and Daw (2018).

4. Rollwage et al. (2020).

5. Klayman (1995); Park et al. (2010); Sunstein et al. (2016); Kappes et al. (2020).

6. Rollwage et al. (2020); Talluri et al. (2018).

7. Rollwage and Fleming (in press).

8. Rollwage, Dolan, and Fleming (2018).

9. De Martino et al. (2013).

10. De Martino et al. (2013); Folke et al. (2016). Intriguingly, people's eye movements also revealed the difficulty of the choice; they looked back and forth between the two options more often when they were uncertain. But only explicit confidence ratings predicted future changes of mind.

11. Frederick (2005); Evans and Stanovich (2013); Thompson et al. (2013); Ackerman and Thompson (2017).

12. Toplak, West, and Stanovich (2011); Pennycook, Fugelsang, and Koehler (2015); Pennycook and Rand (2019); Young and Shtulman (2020).

13. Johnson and Fowler (2011).

14. Anderson et al. (2012).

15. Hertz et al. (2017); Von Hippel and Trivers (2011).

16. Bang et al. (2017); Bang and Fleming (2018); Bang et al. (2020).

17. Edelson et al. (2018); Fleming and Bang (2018); Dalio (2017).

18. Amazon, "To Our Shareholders," 2017, www.sec.gov/Archives/edgar /data/1018724/000119312518121161/d456916dex991.htm.

Chapter 8: Collaborating and Sharing

1. Shea et al. (2014); Frith (2012).

2. This is a version of the Bayesian algorithm that we encountered in Chapter 1 that combines two sensory modalities, such as seeing and hearing, except now the combination is being done across brains, rather than within the same brain. The model also suggests that the benefit of interaction breaks down if the observers are too dissimilar. This is indeed what was found: for dissimilar pairs, two heads are worse than the better one. Bahrami et al. (2010); Fusaroli et al. (2012); Bang et al. (2014); Koriat (2012).

3. Bang et al. (2017); Patel, Fleming, and Kilner (2012); Jiang and Pell (2017); Jiang, Gossack-Keenan, and Pell (2020); Goupil et al. (2020); Pulford et al. (2018).

4. Brewer and Burke (2002); Fox and Walters (1986).

5. Busey et al. (2000).

6. Wixted and Wells (2017).

7. National Research Council (2015).

8. Barney Thompson, "'Reasonable Prospect' of Lawyer Being Vague on Case's Chances," *Financial Times*, November 25, 2018, www.ft.com/content

/94cddbe8-ef31-11e8-8180-9cf212677a57; Robert Rothkopf, "Part 1: Miscommunication in Legal Advice," Balance Legal Capital, November 23, 2018, www.balancelegalcapital.com/litigation-superforecasting-miscommunication/.

9. Tetlock and Gardner (2016).

10. Firestein (2012).

11. Open Science Collaboration (2015); Camerer et al. (2018).

12. Rohrer et al. (in press).

13. Fetterman and Sassenberg (2015).

14. Camerer et al. (2018).

15. Pennycook and Rand (2019).

16. Rollwage, Dolan, and Fleming (2018); Ortoleva and Snowberg (2015); Toner et al. (2013).

17. Schulz et al. (2020).

18. Fischer, Amelung, and Said (2019).

19. Leary et al. (2017).

20. Bang and Frith (2017); Tetlock and Gardner (2016).

Chapter 9: Explaining Ourselves

1. Cleeremans (2011); Norman and Shallice (1986).

2. Beilock and Carr (2001); Beilock et al. (2002).

3. Sian Beilock, "The Best Players Rarely Make the Best Coaches," *Psychology Today*, August 16, 2010, www.psychologytoday.com/intl/blog/choke/201008/the-best-players-rarely-make-the-best-coaches; Steven Rynne and Chris Cushion, "Playing Is Not Coaching: Why So Many Sporting Greats Struggle as Coaches," *The Conversation*, February 8, 2017, http://theconversation.com/playing-is-not-coaching-why-so-many-sporting-greats-struggle-as-coaches-71625.

4. Quoted in James McWilliams, "The Lucrative Art of Chicken Sexing," *Pacific Standard*, September 8, 2018, https://psmag.com/magazine/the-lucrative-art-of-chicken-sexing. It is plausible that global metacognition remains intact in these cases. Despite having poor local insight into choice accuracy, the expert sexers presumably know they are experts.

5. Weiskrantz et al. (1974); Ro et al. (2004); Azzopardi and Cowey (1997); Schmid et al. (2010); Ajina et al. (2015). See Phillips (2020) for the view that blindsight is equivalent to degraded conscious vision, rather than qualitatively unconscious.

6. Kentridge and Heywood (2000); Persaud, McLeod, and Cowey (2007); Ko and Lau (2012).

7. Hall et al. (2010).

8. Johansson et al. (2005); Nisbett and Wilson (1977). The Lund experiments build on a famous paper published in 1977 by the social psychologists Richard Nisbett and Timothy Wilson. The article, "Telling More Than We

Can Know: Verbal Reports on Mental Processes," is one of the most cited in psychology. They surveyed literature suggesting that we have little insight into the mental processes that guide our behavior. This was in part based on experiments in which individuals would have a pronounced choice bias—for instance, choosing the rightmost item in a set of otherwise identical clothing—without any inkling that they were biased in such a fashion.

9. Gazzaniga (1998); Gazzaniga (2005).

10. Hirstein (2005); Benson et al. (1996); Stuss et al. (1978).

11. Wegner (2003).

12. Wegner and Wheatley (1999); Moore et al. (2009); Metcalfe et al. (2012); Wenke, Fleming, and Haggard (2010).

13. Vygotsky (1986); Fernyhough (2016).

14. Schechtman (1996); Walker (2012).

15. Frankfurt (1988); Dennett (1988); Moeller and Goldstein (2014).

16. Crockett et al. (2013); Bulley and Schacter (2020).

17. Pittampalli (2016).

18. Stephen M. Fleming, "Was It Really Me?," *Aeon*, September 26, 2012, https://aeon.co/essays/will-neuroscience-overturn-the-criminal-law-system.

19. Case (2016); Keene et al. (2019).

Chapter 10: Self-Awareness in the Age of Machines

1. Hamilton, Cairns, and Cooper (1961).

2. Turing (1937); Domingos (2015); Tegmark (2017); Marcus and Davis (2019).

3. Braitenberg (1984).

4. Rosenblatt (1958); Rumelhart, Hinton, and Williams (1986).

5. Bengio (2009); Krizhevsky, Sutskever, and Hinton (2012); Schäfer and Zimmermann (2007).

6. Some caveats apply here. What does and does not count as an internal representation is a contentious issue in cognitive science. The philosopher Nicholas Shea has suggested that representations come into play when they help explain the functions of what the system is doing beyond simple, causal chains. For instance, we might say that part of a neural network or a brain region "represents" seeing my son's face, because that representation is activated under many different conditions and many different inputs—it will become active regardless of whether I see his face from the side or from the front, or whether the lighting conditions are good or poor—and that such a representation is useful as it leads me to act in a particular, caring way toward him. In more technical terms, representations like these are "invariant," or robust, across multiple conditions, which makes them a useful level of explanation in psychology and cognitive science (imagine trying to explain what face-selective neurons in a neural network were doing without talking

about faces, and instead just talking about light and shade and inputs and outputs). Pitt (2018); Shea (2018).

7. Khaligh-Razavi and Kriegeskorte (2014); Kriegeskorte (2015); Güçlü and Van Gerven (2015).

8. Silver et al. (2017).

9. James (1950).

10. Marcus and Davis (2019).

11. Clark and Karmiloff-Smith (1993); Cleeremans (2014).

12. Yeung, Cohen, and Botvinick (2004).

13. Pasquali, Timmermans, and Cleeremans (2010). For other related examples, see Insabato et al. (2010) and Atiya et al. (2019).

14. Daftry et al. (2016); Dequaire et al. (2016); Gurău et al. (2018); Gal and Ghahramani (2016); Kendall and Gal (2017).

15. Clark and Karmiloff-Smith (1993); Wang et al. (2018).

16. Georgopoulos et al. (1982).

17. Hochberg et al. (2006).

18. Schurger et al. (2017).

19. Rouault et al. (2018).

20. Samek et al. (2019).

21. Harari (2018).

Chapter 11: Emulating Socrates

1. Maguire et al. (2000); Schneider et al. (2002); Zatorre, Fields, and Johansen-Berg (2012); Draganski et al. (2004); Woollett and Maguire (2011); Scholz et al. (2009); Lerch et al. (2011).

2. Harty et al. (2014); Shekhar and Rahnev (2018).

3. Hester et al. (2012); Hauser et al. (2017); Joensson et al. (2015).

4. Cortese et al. (2016); Cortese et al. (2017).

5. Carpenter et al. (2019).

6. Sinclair, Stanley, and Seli (2020); De Beukelaer et al., (2021).

7. Cosentino (2014); Leary (2007).

8. This intuition has received support in the lab. Careful experiments show that subjects can only remember and report a subset of several objects (such as letters or a bunch of oriented lines) that are briefly flashed on a screen. This itself is not surprising, because our ability to keep multiple things in mind at any one time is limited. Strikingly, however, subjects are able to accurately identify any individual object in the array if they are asked to do so immediately after the set of objects has disappeared. The implication is that all the objects in the array were consciously seen at the time of the stimulus, but that without being cued to report individual locations, they are quickly forgotten and so overflow our ability to report them. Sperling (1960); Landman, Spekreijse, and Lamme (2003); Block (2011); Bronfman et al. (2014).

9. Phillips (2018); Stazicker (2018); Cova, Gaillard, and Kammerer (2020).

10. Lau and Rosenthal (2011); Rosenthal (2005); Brown, Lau, and LeDoux (2019).

11. Lau and Passingham (2006); Michel and Morales (2020); Panagio-taropoulos (2012).

12. Del Cul et al. (2009); Fleming et al. (2014); Sahraie et al. (1997); Persaud et al. (2011).

13. Chalmers (2018); Graziano (2019).

14. Dehaene, Lau, and Kouider (2017); Schooler (2002); Winkielman and Schooler (2009).

15. LeDoux (2016); LeDoux and Brown (2017).

16. La Berge et al. (1981); Baird et al. (2018); Dresler et al. (2012); Voss et al. (2014); Filevich et al. (2015).

17. Baird et al. (2014); Fox et al. (2012).

18. Schmidt et al. (2019); Van Dam et al. (2018); Di Stefano et al. (2016).

REFERENCES

Ackerman, Rakefet, and Morris Goldsmith. "Metacognitive Regulation of Text Learning: On Screen Versus on Paper." *Journal of Experimental Psychology: Applied* 17, no. 1 (2011): 18.

Ackerman, Rakefet, and Valerie A. Thompson. "Meta-Reasoning: Monitoring and Control of Thinking and Reasoning." *Trends in Cognitive Sciences* 21, no. 8 (2017): 607–617.

Adams, Rick A., Stewart Shipp, and Karl J. Friston. "Predictions Not Commands: Active Inference in the Motor System." *Brain Structure & Function* 218, no. 3 (2013): 611–643.

Agnew, Sarah Kathleen, and R. G. Morris. "The Heterogeneity of Anosognosia for Memory Impairment in Alzheimer's Disease: A Review of the Literature and a Proposed Model." *Aging & Mental Health* 2, no. 1 (1998): 7–19.

Ais, Joaquín, Ariel Zylberberg, Pablo Barttfeld, and Mariano Sigman. "Individual Consistency in the Accuracy and Distribution of Confidence Judgments." *Cognition* 146 (2016): 377–386.

Ajina, Sara, Franco Pestilli, Ariel Rokem, Christopher Kennard, and Holly Bridge. "Human Blindsight Is Mediated by an Intact Geniculo-Extrastriate Pathway." *eLife* 4 (2015): e08935.

Alicke, Mark D., Mary L. Klotz, David L. Breitenbecher, Tricia J. Yurak, and Debbie S. Vredenburg. "Personal Contact, Individuation, and the Better-Than-Average Effect." *Journal of Personality and Social Psychology* 68, no. 5 (1995): 804.

Allen, Micah, Darya Frank, D. Samuel Schwarzkopf, Francesca Fardo, Joel S. Winston, Tobias U. Hauser, and Geraint Rees. "Unexpected Arousal Modulates the Influence of Sensory Noise on Confidence." *eLife* 5 (2016): 403.

Allen, Micah, James C. Glen, Daniel Müllensiefen, Dietrich Samuel Schwarz-kopf, Francesca Fardo, Darya Frank, Martina F. Callaghan, and Geraint Rees. "Metacognitive Ability Correlates with Hippocampal and Prefrontal Microstructure." *NeuroImage* 149 (2017): 415–423.

Alter, Adam L., and Daniel M. Oppenheimer. "Predicting Short-Term Stock Fluctuations by Using Processing Fluency." *Proceedings of the National Academy of Sciences* 103, no. 24 (2006): 9369–9372.

———. "Uniting the Tribes of Fluency to Form a Metacognitive Nation." *Personality and Social Psychology Review* 13, no. 3 (2009): 219–235.

Amodio, D. M., and Chris D. Frith. "Meeting of Minds: The Medial Frontal Cortex and Social Cognition." *Nature Reviews Neuroscience* 7, no. 4 (2006): 268–277.

Anderson, Cameron, Sebastien Brion, Don A. Moore, and Jessica A. Kennedy. "A Status-Enhancement Account of Overconfidence." *Journal of Personality and Social Psychology* 103, no. 4 (2012): 718–735.

Apps, Matthew A. J., and Manos Tsakiris. "The Free-Energy Self: A Predictive Coding Account of Self-Recognition." *Neuroscience & Biobehavioral Reviews* 41 (2014): 85–97.

Atiya, Nadim A. A., Iñaki Rañó, Girijesh Prasad, and KongFatt Wong-Lin. "A Neural Circuit Model of Decision Uncertainty and Change-of-Mind." *Nature Communications* 10, no. 1 (2019): 2287.

Aubert, Maxime, Adam Brumm, Muhammad Ramli, Thomas Sutikna, E. Wahyu Saptomo, Budianto Hakim, Michael J. Morwood, Gerrit D. van den Bergh, Leslie Kinsley, and Anthony Dosseto. "Pleistocene Cave Art from Sulawesi, Indonesia." *Nature* 514, no. 7521 (2014): 223–227.

Azzopardi, Paul, and Alan Cowey. "Is Blindsight Like Normal, Near-Threshold Vision?" *Proceedings of the National Academy of Sciences* 94, no. 25 (1997): 14190–14194.

Badre, David, and Derek Evan Nee. "Frontal Cortex and the Hierarchical Control of Behavior." *Trends in Cognitive Sciences* 22, no. 2 (2018): 170–188.

Baggini, Julian. *How the World Thinks: A Global History of Philosophy.* 2018. Reprint, London: Granta, 2019.

Bahrami, Bahador, Karsten Olsen, Peter E. Latham, Andreas Roepstorff, Geraint Rees, and Chris D. Frith. "Optimally Interacting Minds." *Science* 329, no. 5995 (2010): 1081–1085.

Baird, Benjamin, Anna Castelnovo, Olivia Gosseries, and Giulio Tononi. "Frequent Lucid Dreaming Associated with Increased Functional Connectivity Between Frontopolar Cortex and Temporoparietal Association Areas." *Scientific Reports* 8, no. 1 (2018): 1–15.

Baird, Benjamin, Matthew Cieslak, Jonathan Smallwood, Scott T. Grafton, and Jonathan W. Schooler. "Regional White Matter Variation Associated with Domain-Specific Metacognitive Accuracy." *Journal of Cognitive Neuroscience* 27, no. 3 (2015): 440–452.

Baird, Benjamin, Michael D. Mrazek, Dawa T. Phillips, and Jonathan W. Schooler. "Domain-Specific Enhancement of Metacognitive Ability Following Meditation Training." *Journal of Experimental Psychology: General* 143, no. 5 (2014): 1972.

Baird, Benjamin, Jonathan Smallwood, Krzysztof J. Gorgolewski, and Daniel S. Margulies. "Medial and Lateral Networks in Anterior Prefrontal Cortex Support Metacognitive Ability for Memory and Perception." *Journal of Neuroscience* 33, no. 42 (2013): 16657–16665.

Bandura, A. "Self-Efficacy: Toward a Unifying Theory of Behavioral Change." *Psychological Review* 84, no. 2 (1977): 191–215.

Bang, Dan, Laurence Aitchison, Rani Moran, Santiago Herce Castanon, Banafsheh Rafiee, Ali Mahmoodi, Jennifer Y. F. Lau, Peter E. Latham, Bahador Bahrami, and Christopher Summerfield. "Confidence Matching in Group Decision-Making." *Nature Human Behaviour* 1, no. 6 (2017): 1–7.

Bang, Dan, Sara Ershadmanesh, Hamed Nili, and Stephen M. Fleming. "Private-Public Mappings in Human Prefrontal Cortex." *eLife* 9 (2020): e56477.

Bang, Dan, and Stephen M. Fleming. "Distinct Encoding of Decision Confidence in Human Medial Prefrontal Cortex." *Proceedings of the National Academy of Sciences* 115, no. 23 (2018): 6082–6087.

Bang, Dan, and Chris D. Frith. "Making Better Decisions in Groups." *Royal Society Open Science* 4, no. 8 (2017): 170193.

Bang, Dan, Riccardo Fusaroli, Kristian Tylén, Karsten Olsen, Peter E. Latham, Jennifer Y. F. Lau, Andreas Roepstorff, Geraint Rees, Chris D. Frith, and Bahador Bahrami. "Does Interaction Matter? Testing Whether a Confidence Heuristic Can Replace Interaction in Collective Decision-Making." *Consciousness and Cognition* 26 (2014): 13–23.

Bargh, John A., and Yaacov Schul. "On the Cognitive Benefits of Teaching." *Journal of Educational Psychology* 72, no. 5 (1980): 593–604.

Baron-Cohen, Simon, Alan M. Leslie, and Uta Frith. "Does the Autistic Child Have a 'Theory of Mind'?" *Cognition* 21, no. 1 (1985): 37–46.

Barttfeld, Pablo, Bruno Wicker, Phil McAleer, Pascal Belin, Yann Cojan, Martin Graziano, Ramón Leiguarda, and Mariano Sigman. "Distinct Patterns of Functional Brain Connectivity Correlate with Objective Performance and Subjective Beliefs." *Proceedings of the National Academy of Sciences* 110, no. 28 (2013): 11577–11582.

Beck, Brianna, Valentina Peña-Vivas, Stephen Fleming, and Patrick Haggard. "Metacognition Across Sensory Modalities: Vision, Warmth, and Nociceptive Pain." *Cognition* 186 (2019): 32–41.

Beilock, Sian L., and Thomas H. Carr. "On the Fragility of Skilled Performance: What Governs Choking Under Pressure?" *Journal of Experimental Psychology: General* 130, no. 4 (2001): 701–725.

Beilock, Sian L., Thomas H. Carr, Clare MacMahon, and Janet L. Starkes. "When Paying Attention Becomes Counterproductive: Impact of Divided

Versus Skill-Focused Attention on Novice and Experienced Performance of Sensorimotor Skills." *Journal of Experimental Psychology: Applied* 8, no. 1 (2002): 6–16.

Bengio, Yoshua. "Learning Deep Architectures for AI." *Foundations and Trends in Machine Learning* 2, no. 1 (2009): 1–127.

Bengtsson, Sara L., Raymond J. Dolan, and Richard E. Passingham. "Priming for Self-Esteem Influences the Monitoring of One's Own Performance." *Social Cognitive and Affective Neuroscience*, June 15, 2010.

Benson, D. F., A. Djenderedjian, B. L. Miller, N. A. Pachana, L. Chang, L. Itti, and I. Mena. "Neural Basis of Confabulation." *Neurology* 46, no. 5 (1996): 1239–1243.

Bentall, Richard P., Guy A. Baker, and Sue Havers. "Reality Monitoring and Psychotic Hallucinations." *British Journal of Clinical Psychology* 30, no. 3 (1991): 213–222.

Beran, Michael J., J. David Smith, Mariana V. C. Coutinho, Justin J. Couchman, and Joseph Boomer. "The Psychological Organization of 'Uncertainty' Responses and 'Middle' Responses: A Dissociation in Capuchin Monkeys (Cebus Apella)." *Journal of Experimental Psychology: Animal Behavior Processes* 35, no. 3 (2009): 371–381.

Bertelson, P. "Ventriloquism: A Case of Crossmodal Perceptual Grouping." *Advances in Psychology* 129 (1999): 347–362.

Bjork, Robert A., John Dunlosky, and Nate Kornell. "Self-Regulated Learning: Beliefs, Techniques, and Illusions." *Annual Review of Psychology* 64, no. 1 (2013): 417–444.

Blakemore, Sarah-Jayne. *Inventing Ourselves: The Secret Life of the Teenage Brain*. London: Doubleday, 2018.

Blakemore, Sarah-Jayne, Daniel Wolpert, and Chris D. Frith. "Why Can't You Tickle Yourself?" *NeuroReport* 11, no. 11 (2000): R11–R16.

Blanke, Olaf, Mel Slater, and Andrea Serino. "Behavioral, Neural, and Computational Principles of Bodily Self-Consciousness." *Neuron* 88, no. 1 (2015): 145–166.

Block, Ned. "Perceptual Consciousness Overflows Cognitive Access." *Trends in Cognitive Sciences* 15, no. 12 (2011): 567–575.

Bocanegra, Bruno R., Fenna H. Poletiek, Bouchra Ftitache, and Andy Clark. "Intelligent Problem-Solvers Externalize Cognitive Operations." *Nature Human Behaviour* 3, no. 2 (2019): 136–142.

Born, Richard T., and David C. Bradley. "Structure and Function of Visual Area MT." *Annual Review of Neuroscience* 28 (2005): 157–189.

Botvinick, Matthew M., and Johnathan Cohen. "Rubber Hands 'Feel' Touch That Eyes See." *Nature* 391, no. 6669 (1998): 756.

Botvinick, Matthew M., Yael Niv, and Andew G. Barto. "Hierarchically Organized Behavior and Its Neural Foundations: A Reinforcement Learning Perspective." *Cognition* 113, no. 3 (2009): 262–280.

Braitenberg, Valentino. *Vehicles: Experiments in Synthetic Psychology*. Cambridge, MA: MIT Press, 1984.

Bretherton, Inge, and Marjorie Beeghly. "Talking About Internal States: The Acquisition of an Explicit Theory of Mind." *Developmental Psychology* 18, no. 6 (1982): 906–921.

Brewer, Neil, and Anne Burke. "Effects of Testimonial Inconsistencies and Eyewitness Confidence on Mock-Juror Judgments." *Law and Human Behavior* 26, no. 3 (2002): 353–364.

Bronfman, Zohar Z., Noam Brezis, Hilla Jacobson, and Marius Usher. "We See More Than We Can Report: 'Cost Free' Color Phenomenality Outside Focal Attention." *Psychological Science* 25, no. 7 (2014): 1394–1403.

Brown, Richard, Hakwan Lau, and Joseph E. LeDoux. "Understanding the Higher-Order Approach to Consciousness." *Trends in Cognitive Sciences* 23, no. 9 (2019): 754–768.

Bulley, Adam, Thomas McCarthy, Sam J. Gilbert, Thomas Suddendorf, and Jonathan Redshaw. "Children Devise and Selectively Use Tools to Offload Cognition." *Current Biology* 30, no. 17 (2020): 3457–3464.

Bulley, Adam, and Daniel L. Schacter. "Deliberating Trade-Offs with the Future." *Nature Human Behaviour* 4, no. 3 (2020): 238–247.

Burgess, P. W., N. Alderman, J. Evans, H. Emslie, and B. A. Wilson. "The Ecological Validity of Tests of Executive Function." *Journal of the International Neuropsychological Society* 4, no. 6 (1998): 547–558.

Busey, T. A., J. Tunnicliff, G. R. Loftus, and E. F. Loftus. "Accounts of the Confidence-Accuracy Relation in Recognition Memory." *Psychonomic Bulletin & Review* 7, no. 1 (2000): 26–48.

Butterfill, Stephen A., and Ian A. Apperly. "How to Construct a Minimal Theory of Mind." *Mind & Language* 28, no. 5 (2013): 606–637.

Call, Josep. "Seeking Information in Non-Human Animals: Weaving a Metacognitive Web." In *Foundations of Metacognition*, edited by Michael J. Beran, Johannes Brandl, Josef Perner, and Joëlle Proust, 62–75. Oxford: Oxford University Press, 2012.

Call, Josep, and Michael Tomasello. "Does the Chimpanzee Have a Theory of Mind? 30 Years Later." *Trends in Cognitive Sciences* 12, no. 5 (2008): 187–192.

Camerer, Colin F., Anna Dreber, Felix Holzmeister, Teck-Hua Ho, Jürgen Huber, Magnus Johannesson, Michael Kirchler, et al. "Evaluating the Replicability of Social Science Experiments in Nature and Science Between 2010 and 2015." *Nature Human Behaviour* 2, no. 9 (2018): 637–644.

Carpenter, Jason, Maxine T. Sherman, Rogier A. Kievit, Anil K. Seth, Hakwan Lau, and Stephen M. Fleming. "Domain-General Enhancements of Metacognitive Ability Through Adaptive Training." *Journal of Experimental Psychology: General* 148, no. 1 (2019): 51–64.

Carruthers, Peter. "How We Know Our Own Minds: The Relationship Between Mindreading and Metacognition." *Behavioral and Brain Sciences* 32, no. 2 (2009): 121–138.

———. "Meta-Cognition in Animals: A Skeptical Look." *Mind & Language* 23, no. 1 (2008): 58–89.

——. *The Opacity of Mind: An Integrative Theory of Self-Knowledge*. New York: Oxford University Press, 2011.

Case, Paula. "Dangerous Liaisons? Psychiatry and Law in the Court of Protection—Expert Discourses of 'Insight' (and 'Compliance')." *Medical Law Review* 24, no. 3 (2016): 360–378.

Cervone, Daniel. "Effects of Envisioning Future Activities on Self-Efficacy Judgments and Motivation: An Availability Heuristic Interpretation." *Cognitive Therapy and Research* 13, no. 3 (1989): 247–261.

Cervone, Daniel, and Philip K. Peake. "Anchoring, Efficacy, and Action: The Influence of Judgmental Heuristics on Self-Efficacy Judgments and Behavior." *Journal of Personality and Social Psychology* 50, no. 3 (1986): 492.

Cesarini, David, Magnus Johannesson, Paul Lichtenstein, and Björn Wallace. "Heritability of Overconfidence." *Journal of the European Economic Association* 7, nos. 2–3 (2009): 617–627.

Chalmers, David J. "The Meta-Problem of Consciousness." *Journal of Consciousness Studies* 25, nos. 9–10 (2018): 6–61.

Chamorro-Premuzic, Tomas, Nicole Harlaar, Corina U. Greven, and Robert Plomin. "More than Just IQ: A Longitudinal Examination of Self-Perceived Abilities as Predictors of Academic Performance in a Large Sample of UK Twins." *Intelligence* 38, no. 4 (2010): 385–392.

Chang, Liangtang, Shikun Zhang, Mu-Ming Poo, and Neng Gong. "Spontaneous Expression of Mirror Self-Recognition in Monkeys After Learning Precise Visual-Proprioceptive Association for Mirror Images." *Proceedings of the National Academy of Sciences* 114, no. 12 (2017): 3258–3263.

Charles, Lucie, Jean-Rémi King, and Stanislas Dehaene. "Decoding the Dynamics of Action, Intention, and Error Detection for Conscious and Subliminal Stimuli." *Journal of Neuroscience* 34, no. 4 (2014): 1158–1170.

Chen, Patricia, Omar Chavez, Desmond C. Ong, and Brenda Gunderson. "Strategic Resource Use for Learning: A Self-Administered Intervention That Guides Self-Reflection on Effective Resource Use Enhances Academic Performance." *Psychological Science* 28, no. 6 (2017): 774–785.

Christoff, Kalina, Alan M. Gordon, Jonathan Smallwood, Rachelle Smith, and Jonathan W. Schooler. "Experience Sampling During fMRI Reveals Default Network and Executive System Contributions to Mind Wandering." *Proceedings of the National Academy of Sciences* 106, no. 21 (2009): 8719–8724.

Clark, Andy. *Supersizing the Mind: Embodiment, Action, and Cognitive Extension*. New York: Oxford University Press, 2010.

——. *Surfing Uncertainty: Prediction, Action, and the Embodied Mind*. New York: Oxford University Press, 2016.

————. "Whatever Next? Predictive Brains, Situated Agents, and the Future of Cognitive Science." *Behavioral and Brain Sciences* 36, no. 3 (2013): 181–204.

Clark, Andy, and David J. Chalmers. "The Extended Mind." *Analysis* 58, no. 1 (1998): 7–19.

Clark, Andy, and Annette Karmiloff-Smith. "The Cognizer's Innards: A Psychological and Philosophical Perspective on the Development of Thought." *Mind & Language* 8, no. 4 (1993): 487–519.

Clarke, F. R., T. G. Birdsall, and W. P. Tanner. "Two Types of ROC Curves and Definition of Parameters." *Journal of the Acoustical Society of America* 31 (1959): 629–630.

Cleeremans, Axel. "Connecting Conscious and Unconscious Processing." *Cognitive Science* 38, no. 6 (2014): 1286–1315.

————. "The Radical Plasticity Thesis: How the Brain Learns to Be Conscious." *Frontiers in Psychology* 2 (2011): 1–12.

Comte, Auguste. *Introduction to Positive Philosophy*. Indianapolis, IN: Hackett Publishing, 1988.

Conant, Roger C., and W. Ross Ashby. "Every Good Regulator of a System Must Be a Model of That System." *International Journal of Systems Science* 1, no. 2 (1970): 89–97.

Cortese, Aurelio, Kaoru Amano, Ai Koizumi, Mitsuo Kawato, and Hakwan Lau. "Multivoxel Neurofeedback Selectively Modulates Confidence Without Changing Perceptual Performance." *Nature Communications* 7 (2016): 13669.

Cortese, Aurelio, Kaoru Amano, Ai Koizumi, Hakwan Lau, and Mitsuo Kawato. "Decoded fMRI Neurofeedback Can Induce Bidirectional Confidence Changes Within Single Participants." *NeuroImage* 149 (2017): 323–337.

Cosentino, Stephanie. "Metacognition in Alzheimer's Disease." In *The Cognitive Neuroscience of Metacognition*, edited by Stephen M. Fleming and Chris D. Frith, 389–407. Cham, Switzerland: Springer, 2014.

Cosentino, Stephanie, Janet Metcalfe, Brady Butterfield, and Yaakov Stern. "Objective Metamemory Testing Captures Awareness of Deficit in Alzheimer's Disease." *Cortex* 43, no. 7 (2007): 1004–1019.

Costall, Alan. "'Introspectionism' and the Mythical Origins of Scientific Psychology." *Consciousness and Cognition* 15, no. 4 (2006): 634–654.

Courage, Mary L., Shannon C. Edison, and Mark L. Howe. "Variability in the Early Development of Visual Self-Recognition." *Infant Behavior and Development* 27, no. 4 (2004): 509–532.

Cova, Florian, Maxence Gaillard, and François Kammerer. "Is the Phenomenological Overflow Argument Really Supported by Subjective Reports?" *Mind & Language*, April 21, 2020 (epub ahead of print).

Craik, Kenneth. *The Nature of Explanation*. 1943. New Impression edition, Cambridge: Cambridge University Press, 1963.

Crockett, Molly J., Barbara R. Braams, Luke Clark, Philippe N. Tobler, Trevor W. Robbins, and Tobias Kalenscher. "Restricting Temptations: Neural Mechanisms of Precommitment." *Neuron* 79, no. 2 (2013): 391–401.

Cross, K. Patricia. "Not Can, but *Will* College Teaching Be Improved?" *New Directions for Higher Education* 1977, no. 17 (1977): 1–15.

Crump, Matthew J. C., and Gordon D. Logan. "Warning: This Keyboard Will Deconstruct—the Role of the Keyboard in Skilled Typewriting." *Psychonomic Bulletin & Review* 17, no. 3 (2010): 394–399.

Csibra, Gergely, and György Gergely. "Natural Pedagogy." *Trends in Cognitive Sciences* 13, no. 4 (2009): 148–153.

Daftry, Shreyansh, Sam Zeng, J. Andrew Bagnell, and Martial Hebert. "Introspective Perception: Learning to Predict Failures in Vision Systems," in 2016 IEEE/RSJ International Conference on Intelligent Robots and Systems (IROS), 1743–1750.

Dalio, Ray. *Principles: Life and Work*. New York: Simon & Schuster, 2017.

Darwin, Charles. *The Expression of the Emotions in Man and Animals*. London: John Murray, 1872.

David, Anthony S., Nicholas Bedford, Ben Wiffen, and James Gilleen. "Failures of Metacognition and Lack of Insight in Neuropsychiatric Disorders." *Philosophical Transactions of the Royal Society B: Biological Sciences* 367, no. 1594 (2012): 1379–1390.

Dawkins, Richard, and Yan Wong. *The Ancestor's Tale: A Pilgrimage to the Dawn of Life*. London: Weidenfeld & Nicolson, 2016.

De Beukelaer, Sophie, Neza Vehar, Max Rollwage, Stephen M. Fleming, and Manos Tsakiris. "Changing Minds About Climate Change: A Pervasive Role for Domain-General Metacognition." *PsyArXiv*, October 21, 2021.

De Gardelle, Vincent, and Pascal Mamassian. "Does Confidence Use a Common Currency Across Two Visual Tasks?" *Psychological Science* 25, no. 6 (2014): 1286–1288.

De Martino, Benedetto, Stephen M. Fleming, Neil Garrett, and Raymond J. Dolan. "Confidence in Value-Based Choice." *Nature Neuroscience* 16, no. 1 (2013): 105–110.

Dehaene, Stanislas, Hakwan Lau, and Sid Kouider. "What Is Consciousness, and Could Machines Have It?" *Science* 358, no. 6362 (2017): 486–492.

Dehaene, Stanislas, M. I. Posner, and D. M. Tucker. "Localization of a Neural System for Error Detection and Compensation." *Psychological Science* 5, no. 5 (1994): 303–305.

De Jager, Bernadet, Margo Jansen, and Gerry Reezigt. "The Development of Metacognition in Primary School Learning Environments." *School Effectiveness and School Improvement* 16, no. 2 (2005): 179–196.

Del Cul, A., S. Dehaene, P. Reyes, E. Bravo, and A. Slachevsky. "Causal Role of Prefrontal Cortex in the Threshold for Access to Consciousness." *Brain* 132, no. 9 (2009): 2531.

Della Gatta, Francesco, Francesca Garbarini, Guglielmo Puglisi, Antonella Leonetti, Annamaria Berti, and Paola Borroni. "Decreased Motor Cortex Excitability Mirrors Own Hand Disembodiment During the Rubber Hand Illusion." *eLife* 5 (2016): e14972.

Dennett, Daniel C. "Conditions of Personhood." In *What Is a Person?*, edited by Michael F. Goodman, 145–167. Totowa, NJ: Humana Press, 1988.

———. *From Bacteria to Bach and Back: The Evolution of Minds*. London: Penguin, 2018.

———. *Kinds of Minds: Toward an Understanding of Consciousness*. New York: Basic Books, 1996.

Dequaire, Julie, Chi Hay Tong, Winston Churchill, and Ingmar Posner. "Off the Beaten Track: Predicting Localisation Performance in Visual Teach and Repeat." In *2016 IEEE International Conference on Robotics and Automation (ICRA)*, 795–800. Stockholm, Sweden: IEEE, 2016.

Descartes, René. *Meditations and Other Metaphysical Writings*. London: Penguin, 1998.

Desender, Kobe, Annika Boldt, and Nick Yeung. "Subjective Confidence Predicts Information Seeking in Decision Making." *Psychological Science* 29, no. 5 (2018): 761–778.

Desender, Kobe, Peter Murphy, Annika Boldt, Tom Verguts, and Nick Yeung. "A Postdecisional Neural Marker of Confidence Predicts Information-Seeking in Decision-Making." *Journal of Neuroscience* 39, no. 17 (2019): 3309–3319.

Diedrichsen, Jörn, Yasmin Hashambhoy, Tushar Rane, and Reza Shadmehr. "Neural Correlates of Reach Errors." *Journal of Neuroscience* 25, no. 43 (2005): 9919–9931.

Diemand-Yauman, Connor, Daniel M. Oppenheimer, and Erikka B. Vaughan. "Fortune Favors the Bold (and the Italicized): Effects of Disfluency on Educational Outcomes." *Cognition* 118, no. 1 (2011): 111–115.

Di Stefano, Giada, Francesca Gino, Gary P. Pisano, and Bradley R. Staats. "Making Experience Count: The Role of Reflection in Individual Learning." NOM Unit Working Paper No. 14-093, Harvard Business School, Boston, MA, June 14, 2016. https://papers.ssrn.com/abstract =2414478.

Domingos, Pedro. *The Master Algorithm: How the Quest for the Ultimate Learning Machine Will Remake Our World*. London: Penguin, 2015.

Draganski, Bogdan, Christian Gaser, Volker Busch, Gerhard Schuierer, Ulrich Bogdahn, and Arne May. "Neuroplasticity: Changes in Grey Matter Induced by Training." *Nature* 427, no. 6972 (2004): 311–312.

Dresler, Martin, Renate Wehrle, Victor I. Spoormaker, Stefan P. Koch, Florian Holsboer, Axel Steiger, Hellmuth Obrig, Philipp G. Sämann, and Michael Czisch. "Neural Correlates of Dream Lucidity Obtained from Contrasting Lucid Versus Non-Lucid REM Sleep: A Combined EEG/fMRI Case Study." *Sleep* 35, no. 7 (2012): 1017–1020.

Dunlosky, John, and Keith W. Thiede. "What Makes People Study More? An Evaluation of Factors That Affect Self-Paced Study." *Acta Psychologica* 98, no. 1 (1998): 37–56.

Dunning, David. *Self-Insight: Roadblocks and Detours on the Path to Knowing Thyself*. New York: Psychology Press, 2012.

Edelson, Micah G., Rafael Polania, Christian C. Ruff, Ernst Fehr, and Todd A. Hare. "Computational and Neurobiological Foundations of Leadership Decisions." *Science* 361, no. 6401 (2018).

Ehrlinger, Joyce, Kerri Johnson, Matthew Banner, David Dunning, and Justin Kruger. "Why the Unskilled Are Unaware: Further Explorations of (Absent) Self-Insight Among the Incompetent." *Organizational Behavior and Human Decision Processes* 105, no. 1 (2008): 98–121.

Eichner, Carolin, and Fabrice Berna. "Acceptance and Efficacy of Metacognitive Training (MCT) on Positive Symptoms and Delusions in Patients with Schizophrenia: A Meta-Analysis Taking into Account Important Moderators." *Schizophrenia Bulletin* 42, no. 4 (2016): 952–962.

Ernst, Marc O., and Martin S. Banks. "Humans Integrate Visual and Haptic Information in a Statistically Optimal Fashion." *Nature* 415, no. 6870 (2002): 429–433.

Eskreis-Winkler, Lauren, Katherine L. Milkman, Dena M. Gromet, and Angela L. Duckworth. "A Large-Scale Field Experiment Shows Giving Advice Improves Academic Outcomes for the Advisor." *Proceedings of the National Academy of Sciences* 116, no. 30 (2019): 14808–14810.

Evans, Jonathan St. B. T., and Keith E. Stanovich. "Dual-Process Theories of Higher Cognition: Advancing the Debate." *Perspectives on Psychological Science* 8, no. 3 (2013): 223–241.

Faivre, Nathan, Elisa Filevich, Guillermo Solovey, Simone Kühn, and Olaf Blanke. "Behavioral, Modeling, and Electrophysiological Evidence for Supramodality in Human Metacognition." *Journal of Neuroscience* 38, no. 2 (2018): 263–277.

Fandakova, Yana, Diana Selmeczy, Sarah Leckey, Kevin J. Grimm, Carter Wendelken, Silvia A. Bunge, and Simona Ghetti. "Changes in Ventromedial Prefrontal and Insular Cortex Support the Development of Metamemory from Childhood into Adolescence." *Proceedings of the National Academy of Sciences* 114, no. 29 (2017): 7582–7587.

Felleman, D. J., and D. C. van Essen. "Distributed Hierarchical Processing in the Primate Cerebral Cortex." *Cerebral Cortex* 1, no. 1 (1991): 1–47.

Fernyhough, Charles. *The Voices Within: The History and Science of How We Talk to Ourselves*. New York: Basic Books, 2016.

Fetterman, Adam K., and Kai Sassenberg. "The Reputational Consequences of Failed Replications and Wrongness Admission Among Scientists." *PLOS One* 10, no. 12 (2015): e0143723.

Filevich, Elisa, Martin Dresler, Timothy R. Brick, and Simone Kühn. "Metacognitive Mechanisms Underlying Lucid Dreaming." *Journal of Neuroscience* 35, no. 3 (2015): 1082–1088.

Filevich, Elisa, Caroline Garcia Forlim, Carmen Fehrman, Carina Forster, Markus Paulus, Yee Lee Shing, and Simone Kühn. "I Know That I Know Nothing: Cortical Thickness and Functional Connectivity Underlying Meta-Ignorance Ability in Pre-schoolers." *Developmental Cognitive Neuroscience* 41 (2020): 100738.

Firestein, Stuart. *Ignorance: How It Drives Science.* New York: Oxford University Press, 2012.

Fischer, Helen, Dorothee Amelung, and Nadia Said. "The Accuracy of German Citizens' Confidence in Their Climate Change Knowledge." *Nature Climate Change* 9, no. 10 (2019): 776–780.

Flavell, J. H. "Metacognition and Cognitive Monitoring: A New Area of Cognitive-Developmental Inquiry." *American Psychologist* 34 (1979): 906–911.

Fleming, Stephen M. "Decision Making: Changing Our Minds About Changes of Mind." *eLife* 5 (2016): e14790.

———. "HMeta-d: Hierarchical Bayesian Estimation of Metacognitive Efficiency from Confidence Ratings." *Neuroscience of Consciousness* 2017, no. 1 (2017).

Fleming, Stephen M., and Dan Bang. "Shouldering Responsibility." *Science* 361, no. 6401 (2018): 449–450.

Fleming, Stephen M., and Nathaniel D. Daw. "Self-Evaluation of Decision-Making: A General Bayesian Framework for Metacognitive Computation." *Psychological Review* 124, no. 1 (2017): 91–114.

Fleming, Stephen M., and Raymond J. Dolan. "The Neural Basis of Metacognitive Ability." *Philosophical Transactions of the Royal Society B: Biological Sciences* 367, no. 1594 (2012): 1338–1349.

Fleming, Stephen M., and Chris D. Frith. *The Cognitive Neuroscience of Metacognition.* Cham, Switzerland: Springer, 2014.

Fleming, Stephen M., Josefien Huijgen, and Raymond J. Dolan. "Prefrontal Contributions to Metacognition in Perceptual Decision Making." *Journal of Neuroscience* 32, no. 18 (2012): 6117–6125.

Fleming, Stephen M., and Hakwan Lau. "How to Measure Metacognition." *Frontiers in Human Neuroscience* 8 (2014): 443.

Fleming, Stephen M., Brian Maniscalco, Yoshiaki Ko, Namema Amendi, Tony Ro, and Hakwan Lau. "Action-Specific Disruption of Perceptual Confidence." *Psychological Science* 26, no. 1 (2015): 89–98.

Fleming, Stephen M., Jihye Ryu, John G. Golfinos, and Karen E. Blackmon. "Domain-Specific Impairment in Metacognitive Accuracy Following Anterior Prefrontal Lesions." *Brain* 137, no. 10 (2014): 2811–2822.

Fleming, Stephen M., Elisabeth J. van der Putten, and Nathaniel D. Daw. "Neural Mediators of Changes of Mind About Perceptual Decisions." *Nature Neuroscience* 21 (2018).

Fleming, Stephen M., Rimona S. Weil, Zoltan Nagy, Raymond J. Dolan, and Geraint Rees. "Relating Introspective Accuracy to Individual Differences in Brain Structure." *Science* 329, no. 5998 (2010): 1541–1543.

Folke, Tomas, Catrine Jacobsen, Stephen M. Fleming, and Benedetto De Martino. "Explicit Representation of Confidence Informs Future Value-Based Decisions." *Nature Human Behaviour* 1, no. 1 (2016): 0002.

Fotopoulou, Aikaterini, Anthony Rudd, Paul Holmes, and Michael Kopelman. "Self-Observation Reinstates Motor Awareness in Anosognosia for Hemiplegia." *Neuropsychologia* 47, no. 5 (2009): 1256–1260.

Fourneret, P., and M. Jeannerod. "Limited Conscious Monitoring of Motor Performance in Normal Subjects." *Neuropsychologia* 36, no. 11 (1998): 1133–1140.

Fox, Kieran C. R., Pierre Zakarauskas, Matt Dixon, Melissa Ellamil, Evan Thompson, and Kalina Christoff. "Meditation Experience Predicts Introspective Accuracy." *PLOS One* 7, no. 9 (2012): e45370.

Fox, Steven G., and H. A. Walters. "The Impact of General Versus Specific Expert Testimony and Eyewitness Confidence Upon Mock Juror Judgment." *Law and Human Behavior* 10, no. 3 (1986): 215–228.

Frankfurt, Harry G. "Freedom of the Will and the Concept of a Person." In *What Is a Person?*, edited by Michael F. Goodman, 127–144. Totowa, NJ: Humana Press, 1988.

Frederick, Shane. "Cognitive Reflection and Decision-Making." *Journal of Economic Perspectives* 19, no. 4 (2005): 25–42.

Freud, Sigmund. *The Interpretation of Dreams*. London: Wordsworth Editions, 1997.

Friston, Karl. "The Free-Energy Principle: A Unified Brain Theory?" *Nature Reviews Neuroscience* 11, no. 2 (2010): 127–138.

Friston, Karl, Jean Daunizeau, James Kilner, and Stefan J. Kiebel. "Action and Behavior: A Free-Energy Formulation." *Biological Cybernetics* 102, no. 3 (2010): 227–260.

Frith, Chris D. *The Cognitive Neuropsychology of Schizophrenia*. Hillsdale, NJ: Lawrence Erlbaum Associates, 1992.

———. "The Role of Metacognition in Human Social Interactions." *Philosophical Transactions of the Royal Society B: Biological Sciences* 367, no. 1599 (2012): 2213–2223.

Fu, Zhongzheng, Daw-An J. Wu, Ian Ross, Jeffrey M. Chung, Adam N. Mamelak, Ralph Adolphs, and Ueli Rutishauser. "Single-Neuron Correlates of Error Monitoring and Post-Error Adjustments in Human Medial Frontal Cortex." *Neuron* 101, no. 1 (2019): 165–177.e5.

Fujita, Kazuo, Noriyuki Nakamura, Sumie Iwasaki, and Sota Watanabe. "Are Birds Metacognitive?" In *Foundations of Metacognition*, edited by Michael J. Beran, Johannes Brandl, Josef Perner, and Joëlle Proust, 50–61. Oxford: Oxford University Press, 2012.

Fusaroli, Riccardo, Bahador Bahrami, Karsten Olsen, Andreas Roepstorff, Geraint Rees, Chris D. Frith, and Kristian Tylén. "Coming to Terms: Quantifying the Benefits of Linguistic Coordination." *Psychological Science* 23, no. 8 (2012): 931–939.

Gadagkar, Vikram, Pavel A. Puzerey, Ruidong Chen, Eliza Baird-Daniel, Alexander R. Farhang, and Jesse H. Goldberg. "Dopamine Neurons Encode Performance Error in Singing Birds." *Science* 354, no. 6317 (2016): 1278–1282.

Gajdos, Thibault, Stephen M. Fleming, Marta Saez Garcia, Gabriel Weindel, and Karen Davranche. "Revealing Subthreshold Motor Contributions to Perceptual Confidence." *Neuroscience of Consciousness* 2019, no. 1 (2019): niz001.

Gal, Yarin, and Zoubin Ghahramani. "Dropout as a Bayesian Approximation: Representing Model Uncertainty in Deep Learning," arXiv.org, October 4, 2016.

Galvin, Susan J., John V. Podd, Vit Drga, and John Whitmore. "Type 2 Tasks in the Theory of Signal Detectability: Discrimination Between Correct and Incorrect Decisions." *Psychonomic Bulletin & Review* 10, no. 4 (2003): 843–876.

Garrison, Jane R., Emilio Fernandez-Egea, Rashid Zaman, Mark Agius, and Jon S. Simons. "Reality Monitoring Impairment in Schizophrenia Reflects Specific Prefrontal Cortex Dysfunction." *NeuroImage: Clinical* 14 (2017): 260–268.

Gazzaniga, Michael S. "Forty-Five Years of Split-Brain Research and Still Going Strong." *Nature Reviews Neuroscience* 6, no. 8 (2005): 653–659.
———. *The Mind's Past*. Berkeley: University of California Press, 1998.

Gehring, W. J., B. Goss, M. G. H. Coles, D. E. Meyer, and E. Donchin. "A Neural System for Error Detection and Compensation." *Psychological Science* 4, no. 6 (1993): 385.

Georgopoulos, A. P., J. F. Kalaska, R. Caminiti, and J. T. Massey. "On the Relations Between the Direction of Two-Dimensional Arm Movements and Cell Discharge in Primate Motor Cortex." *Journal of Neuroscience* 2, no. 11 (1982): 1527–1537.

Gershman, Samuel J., and Nathaniel D. Daw. "Perception, Action and Utility: The Tangled Skein." *Principles of Brain Dynamics: Global State Interactions*, 2012, 293–312.

Gherman, Sabina, and Marios Philiastides. "Human VMPFC Encodes Early Signatures of Confidence in Perceptual Decisions." *eLife* 7 (2018).

Gilbert, Sam J. "Strategic Use of Reminders: Influence of Both Domain-General and Task-Specific Metacognitive Confidence, Independent of Objective Memory Ability." *Consciousness and Cognition* 33 (2015): 245–260.

Gold, J. I., and M. N. Shadlen. "Banburismus and the Brain: Decoding the Relationship Between Sensory Stimuli, Decisions, and Reward." *Neuron* 36, no. 2 (2002): 299–308.

Goldberg, Ilan I., Michal Harel, and Rafael Malach. "When the Brain Loses Its Self: Prefrontal Inactivation During Sensorimotor Processing." *Neuron* 50, no. 2 (2006): 329–339.

Gopnik, Alison, and Janet W. Astington. "Children's Understanding of Representational Change and Its Relation to the Understanding of False Belief and the Appearance-Reality Distinction." *Child Development* 59, no. 1 (1988): 26–37.

Goupil, Louise, and Sid Kouider. "Behavioral and Neural Indices of Metacognitive Sensitivity in Preverbal Infants." *Current Biology* 26, no. 22 (2016): 3038–3045.

Goupil, Louise, Emmanuel Ponsot, Daniel C. Richardson, Gabriel Reyes, and Jean-Julien Aucouturier. "Listeners' Perceptions of the Certainty and Honesty of a Speaker Are Associated With a Common Prosodic Signature." *Nature Communications* 12 (2021): 861.

Goupil, Louise, Margaux Romand-Monnier, and Sid Kouider. "Infants Ask for Help When They Know They Don't Know." *Proceedings of the National Academy of Sciences* 113, no. 13 (2016): 3492–3496.

Graziano, Michael S. A. *Rethinking Consciousness: A Scientific Theory of Subjective Experience*. New York: W. W. Norton, 2019.

Green, D. M., and J. A. Swets. *Signal Detection Theory and Psychophysics*. New York: Wiley, 1966.

Gregory, Richard. *The Intelligent Eye*. New York: McGraw-Hill, 1970.

Greven, Corina U., Nicole Harlaar, Yulia Kovas, Tomas Chamorro-Premuzic, and Robert Plomin. "More than Just IQ: School Achievement Is Predicted by Self-Perceived Abilities—but for Genetic Rather than Environmental Reasons." *Psychological Science* 20, no. 6 (2009): 753–762.

Güçlü, Umut, and Marcel A. J. van Gerven. "Deep Neural Networks Reveal a Gradient in the Complexity of Neural Representations Across the Ventral Stream." *Journal of Neuroscience* 35, no. 27 (2015): 10005–10014.

Gurău, Corina, Dushyant Rao, Chi Hay Tong, and Ingmar Posner. "Learn from Experience: Probabilistic Prediction of Perception Performance to Avoid Failure." *International Journal of Robotics Research* 37, no. 9 (2018): 981–995.

Hall, Lars, Petter Johansson, Betty Tärning, Sverker Sikström, and Thérèse Deutgen. "Magic at the Marketplace: Choice Blindness for the Taste of Jam and the Smell of Tea." *Cognition* 117, no. 1 (2010): 54–61.

Hamilton, Edith, Huntington Cairns, and Lane Cooper. *The Collected Dialogues of Plato*. Princeton, NJ: Princeton University Press, 1961.

Hampton, R. R. "Rhesus Monkeys Know When They Remember." *Proceedings of the National Academy of Sciences* 98, no. 9 (2001): 5359–5362.

Harari, Yuval Noah. *21 Lessons for the 21st Century*. London: Jonathan Cape, 2018.

Hart, J. T. "Memory and the Feeling-of-Knowing Experience." *Journal of Educational Psychology* 56, no. 4 (1965): 208–216.

Harty, S., I. H. Robertson, C. Miniussi, O. C. Sheehy, C. A. Devine, S. McCreery, and R. G. O'Connell. "Transcranial Direct Current Stimulation over Right Dorsolateral Prefrontal Cortex Enhances Error Awareness in Older Age." *Journal of Neuroscience* 34, no. 10 (2014): 3646–3652.

Harvey, Philip D. "Reality Monitoring in Mania and Schizophrenia: The Association of Thought Disorder and Performance." *Journal of Nervous and Mental Disease* 173, no. 2 (1985): 67–73.

Hasbroucq, T., C. A. Possamaï, M. Bonnet, and F. Vidal. "Effect of the Irrelevant Location of the Response Signal on Choice Reaction Time: An Electromyographic Study in Humans." *Psychophysiology* 36, no. 4 (1999): 522–526.

Hauser, Tobias U., Micah Allen, Nina Purg, Michael Moutoussis, Geraint Rees, and Raymond J. Dolan. "Noradrenaline Blockade Specifically Enhances Metacognitive Performance." *eLife* 6 (2017): 468.

Helmholtz, H. L. F. *Treatise on Physiological Optics*. London: Thoemmes Continuum, 1856.

Hembacher, Emily, and Simona Ghetti. "Don't Look at My Answer: Subjective Uncertainty Underlies Preschoolers' Exclusion of Their Least Accurate Memories." *Psychological Science* 25, no. 9 (2014): 1768–1776.

Herculano-Houzel, Suzana. *The Human Advantage: A New Understanding of How Our Brain Became Remarkable*. Cambridge, MA: MIT Press, 2016.

Herculano-Houzel, Suzana, Kamilla Avelino-de-Souza, Kleber Neves, Jairo Porfírio, Débora Messeder, Larissa Mattos Feijó, José Maldonado, and Paul R. Manger. "The Elephant Brain in Numbers." *Frontiers in Neuroanatomy* 8 (2014): 46.

Herculano-Houzel, Suzana, Christine E. Collins, Peiyan Wong, and Jon H. Kaas. "Cellular Scaling Rules for Primate Brains." *Proceedings of the National Academy of Sciences* 104, no. 9 (2007): 3562–3567.

Herculano-Houzel, Suzana, Bruno Mota, and Roberto Lent. "Cellular Scaling Rules for Rodent Brains." *Proceedings of the National Academy of Sciences* 103, no. 32 (2006): 12138–12143.

Hertz, Uri, Stefano Palminteri, Silvia Brunetti, Cecilie Olesen, Chris D. Frith, and Bahador Bahrami. "Neural Computations Underpinning the Strategic Management of Influence in Advice Giving." *Nature Communications* 8, no. 1 (2017): 1–12.

Hester, Robert, L. Sanjay Nandam, Redmond G. O'Connell, Joe Wagner, Mark Strudwick, Pradeep J. Nathan, Jason B. Mattingley, and Mark A. Bellgrove. "Neurochemical Enhancement of Conscious Error Awareness." *Journal of Neuroscience* 32, no. 8 (2012): 2619–2627.

Heyes, Cecilia. "Animal Mindreading: What's the Problem?" *Psychonomic Bulletin & Review* 22, no. 2 (2015): 313–327.

———. *Cognitive Gadgets: The Cultural Evolution of Thinking*. Cambridge, MA: Belknap Press, 2018.

———. "Reflections on Self-Recognition in Primates." *Animal Behaviour* 47, no. 4 (1994): 909–919.

Heyes, Cecilia, Dan Bang, Nicholas Shea, Chris D. Frith, and Stephen M. Fleming. "Knowing Ourselves Together: The Cultural Origins of Metacognition." *Trends in Cognitive Sciences* 24, no. 5 (2020): 349–362.

Heyes, Cecilia, and Chris D. Frith. "The Cultural Evolution of Mind Reading." *Science* 344, no. 6190 (2014): 1243091.

Higham, P. A. "No Special K! A Signal Detection Framework for the Strategic Regulation of Memory Accuracy." *Journal of Experimental Psychology: General* 136, no. 1 (2007): 1–22.

Hilgenstock, Raphael, Thomas Weiss, and Otto W. Witte. "You'd Better Think Twice: Post-Decision Perceptual Confidence." *NeuroImage* 99 (2014): 323–331.

Hirstein, William. *Brain Fiction: Self-Deception and the Riddle of Confabulation*. Cambridge, MA: MIT Press, 2005.

Hisey, Erin, Matthew Gene Kearney, and Richard Mooney. "A Common Neural Circuit Mechanism for Internally Guided and Externally Reinforced Forms of Motor Learning." *Nature Neuroscience* 21, no. 4 (2018): 1–13.

Hochberg, Leigh R., Mijail D. Serruya, Gerhard M. Friehs, Jon A. Mukand, Maryam Saleh, Abraham H. Caplan, Almut Branner, David Chen, Richard D. Penn, and John P. Donoghue. "Neuronal Ensemble Control of Prosthetic Devices by a Human with Tetraplegia." *Nature* 442, no. 7099 (2006): 164–171.

Hodges, Andrew. *Alan Turing: The Enigma*. London: Vintage, 1992.

Hohwy, Jakob. *The Predictive Mind*. Oxford: Oxford University Press, 2013.

Hoven, Monja, Maël Lebreton, Jan B. Engelmann, Damiaan Denys, Judy Luigjes, and Ruth J. van Holst. "Abnormalities of Confidence in Psychiatry: An Overview and Future Perspectives." *Translational Psychiatry* 9, no. 1 (2019): 1–18.

Howard, Charlotte E., Pilar Andrés, Paul Broks, Rupert Noad, Martin Sadler, Debbie Coker, and Giuliana Mazzoni. "Memory, Metamemory and Their Dissociation in Temporal Lobe Epilepsy." *Neuropsychologia* 48, no. 4 (2010): 921–932.

Hu, Xiao, Zhaomin Liu, Tongtong Li, and Liang Luo. "Influence of Cue Word Perceptual Information on Metamemory Accuracy in Judgement of Learning." *Memory* 24, no. 3 (2015): 1–16.

Hu, Xiao, Liang Luo, and Stephen M. Fleming. "A Role for Metamemory in Cognitive Offloading." *Cognition* 193 (2019): 104012.

Hughes, Claire, Sara R. Jaffee, Francesca Happé, Alan Taylor, Avshalom Caspi, and Terrie E. Moffitt. "Origins of Individual Differences in Theory of Mind: From Nature to Nurture?" *Child Development* 76, no. 2 (2005): 356–370.

Insabato, Andrea, Mario Pannunzi, and Gustavo Deco. "Neural Correlates of Metacognition: A Critical Perspective on Current Tasks." *Neuroscience & Biobehavioral Reviews* 71 (2016): 167–175.

Insabato, Andrea, Mario Pannunzi, Edmund T. Rolls, and Gustavo Deco. "Confidence-Related Decision Making." *Journal of Neurophysiology* 104, no. 1 (2010): 539–547.

Ivanhoe, Philip J., trans. *The Daodejing of Laozi*. Indianapolis, IN: Hackett Publishing, 2003.

James, William. *The Principles of Psychology*. Vol. 1. Mineola, NY: Dover Publications, 1950.

Janowsky, J. S., Arthur P. Shimamura, M. Kritchevsky, and L. R. Squire. "Cognitive Impairment Following Frontal Lobe Damage and Its Relevance to Human Amnesia." *Behavioral Neuroscience* 103, no. 3 (1989): 548.

Jenkins, Adrianna C., C. Neil Macrae, and Jason P. Mitchell. "Repetition Suppression of Ventromedial Prefrontal Activity During Judgments of Self and Others." *Proceedings of the National Academy of Sciences* 105, no. 11 (2008): 4507–4512.

Jiang, Xiaoming, Kira Gossack-Keenan, and Marc D. Pell. "To Believe or Not to Believe? How Voice and Accent Information in Speech Alter Listener Impressions of Trust." *Quarterly Journal of Experimental Psychology* 73, no. 1 (2020): 55–79.

Jiang, Xiaoming, and Marc D. Pell. "The Sound of Confidence and Doubt." *Speech Communication* 88 (2017): 106–126.

Joensson, Morten, Kristine Rømer Thomsen, Lau M. Andersen, Joachim Gross, Kim Mouridsen, Kristian Sandberg, Leif Østergaard, and Hans C. Lou. "Making Sense: Dopamine Activates Conscious Self-Monitoring Through Medial Prefrontal Cortex." *Human Brain Mapping* 36, no. 5 (2015): 1866–1877.

Johansson, Petter, Lars Hall, Sverker Sikstrom, and Andreas Olsson. "Failure to Detect Mismatches Between Intention and Outcome in a Simple Decision Task." *Science* 310, no. 5745 (2005): 116–119.

Johnson, Dominic D. P., and James H. Fowler. "The Evolution of Overconfidence." *Nature* 477, no. 7364 (2011): 317–320.

Johnson, Marcia K., and Carol L. Raye. "Reality Monitoring." *Psychological Review* 88, no. 1 (1981): 67.

Jonas, Eric, and Konrad Paul Kording. "Could a Neuroscientist Understand a Microprocessor?" *PLOS Computational Biology* 13, no. 1 (2017): e1005268.

Jönsson, Fredrik U., Henrik Olsson, and Mats J. Olsson. "Odor Emotionality Affects the Confidence in Odor Naming." *Chemical Senses* 30, no. 1 (2005): 29–35.

Jordano, Megan L., and Dayna R. Touron. "How Often Are Thoughts Metacognitive? Findings from Research on Self-Regulated Learning, Think-Aloud Protocols, and Mind-Wandering." *Psychonomic Bulletin & Review* 25, no. 4 (2018): 1269–1286.

Kahneman, Daniel. *Thinking, Fast and Slow*. London: Penguin, 2012.

Kaminski, Juliane, Josep Call, and Michael Tomasello. "Chimpanzees Know What Others Know, but Not What They Believe." *Cognition* 109, no. 2 (2008): 224–234.

Kao, Yun-Ching, Emily S. Davis, and John D. E. Gabrieli. "Neural Correlates of Actual and Predicted Memory Formation." *Nature Neuroscience* 8, no. 12 (2005): 1776–1783.

Kappes, Andreas, Ann H. Harvey, Terry Lohrenz, P. Read Montague, and Tali Sharot. "Confirmation Bias in the Utilization of Others' Opinion Strength." *Nature Neuroscience* 23, no. 1 (2020): 130–137.

Karpicke, Jeffrey D. "Metacognitive Control and Strategy Selection: Deciding to Practice Retrieval During Learning." *Journal of Experimental Psychology: General* 138, no. 4 (2009): 469–486.

Kay, Katty, and Claire Shipman. *The Confidence Code: The Science and Art of Self-Assurance—What Women Should Know.* New York: Harper Business, 2014.

Keene, Alex Ruck, Nuala B. Kane, Scott Y. H. Kim, and Gareth S. Owen. "Taking Capacity Seriously? Ten Years of Mental Capacity Disputes Before England's Court of Protection." *International Journal of Law and Psychiatry* 62 (2019): 56–76.

Kelley, W. M., C. N. Macrae, C. L. Wyland, S. Caglar, S. Inati, and T. F. Heatherton. "Finding the Self? An Event-Related fMRI Study." *Journal of Cognitive Neuroscience* 14, no. 5 (2002): 785–794.

Kendall, Alex, and Yarin Gal. "What Uncertainties Do We Need in Bayesian Deep Learning for Computer Vision?" arXiv.org, October 5, 2017.

Kentridge, R. W., and C. A. Heywood. "Metacognition and Awareness." *Consciousness and Cognition* 9, no. 2 (2000): 308–312.

Kepecs, Adam, Naoshige Uchida, Hatim A. Zariwala, and Zachary F. Mainen. "Neural Correlates, Computation and Behavioural Impact of Decision Confidence." *Nature* 455, no. 7210 (2008): 227–231.

Kersten, Daniel, Pascal Mamassian, and Alan Yuille. "Object Perception as Bayesian Inference." *Annual Review of Psychology* 55 (2004): 271–304.

Khaligh-Razavi, Seyed-Mahdi, and Nikolaus Kriegeskorte. "Deep Supervised, but Not Unsupervised, Models May Explain IT Cortical Representation." *PLOS Computational Biology* 10, no. 11 (2014): e1003915.

Kiani, R., and M. N. Shadlen. "Representation of Confidence Associated with a Decision by Neurons in the Parietal Cortex." *Science* 324, no. 5928 (2009): 759–764.

Klayman, Joshua. "Varieties of Confirmation Bias." In *The Psychology of Learning and Motivation.* Vol. 32, edited by Jerome Busemeyer, Reid Hastie, Douglas L. Medin, 385–418. Cambridge, MA: Academic Press, 1995.

Kloo, Daniela, Michael Rohwer, and Josef Perner. "Direct and Indirect Admission of Ignorance by Children." *Journal of Experimental Child Psychology* 159 (2017): 279–295.

Knoblich, Günther, Frank Stottmeister, and Tilo Kircher. "Self-Monitoring in Patients with Schizophrenia." *Psychological Medicine* 34, no. 8 (2004): 1561.

Knoll, Abby R., Hajime Otani, Reid L. Skeel, and K. Roger van Horn. "Learning Style, Judgements of Learning, and Learning of Verbal and Visual Information." *British Journal of Psychology* 108, no. 3 (2017): 544–563.

Ko, Yoshiaki, and Hakwan Lau. "A Detection Theoretic Explanation of Blindsight Suggests a Link Between Conscious Perception and Metacognition." *Philosophical Transactions of the Royal Society B: Biological Sciences* 367, no. 1594 (2012): 1401–1411.

Kohda, Masanori, Takashi Hotta, Tomohiro Takeyama, Satoshi Awata, Hirokazu Tanaka, Jun-ya Asai, and Alex L. Jordan. "If a Fish Can Pass the Mark Test, What Are the Implications for Consciousness and Self-Awareness Testing in Animals?" *PLOS Biology* 17, no. 2 (2019): e3000021.

Koizumi, Ai, Brian Maniscalco, and Hakwan Lau. "Does Perceptual Confidence Facilitate Cognitive Control?" *Attention, Perception and Psychophysics* 77, no. 4 (2015): 1295–1306.

Komura, Yutaka, Akihiko Nikkuni, Noriko Hirashima, Teppei Uetake, and Aki Miyamoto. "Responses of Pulvinar Neurons Reflect a Subject's Confidence in Visual Categorization." *Nature Neuroscience* 16, no. 6 (2013): 746–755.

Koriat, Asher. "When Are Two Heads Better than One and Why?" *Science* 336, no. 6079 (2012): 360–362.

Koriat, Asher, and Rakefet Ackerman. "Metacognition and Mindreading: Judgments of Learning for Self and Other During Self-Paced Study." *Consciousness and Cognition* 19, no. 1 (2010): 251–264.

Koriat, Asher, and M. Goldsmith. "Monitoring and Control Processes in the Strategic Regulation of Memory Accuracy." *Psychological Review* 103, no. 3 (1996): 490–517.

Kornell, Nate. "Optimising Learning Using Flashcards: Spacing Is More Effective than Cramming." *Applied Cognitive Psychology* 23, no. 9 (2009): 1297–1317.

Kornell, Nate, and Janet Metcalfe. "Study Efficacy and the Region of Proximal Learning Framework." *Journal of Experimental Psychology: Learning, Memory and Cognition* 32, no. 3 (2006): 609.

Kornell, Nate, and Lisa K. Son. "Learners' Choices and Beliefs About Self-Testing." *Memory* 17, no. 5 (2009): 493–501.

Kornell, Nate, Lisa K. Son, and H. S. Terrace. "Transfer of Metacognitive Skills and Hint Seeking in Monkeys." *Psychological Science* 18, no. 1 (2007): 64–71.

Kriegeskorte, Nikolaus. "Deep Neural Networks: A New Framework for Modeling Biological Vision and Brain Information Processing." *Annual Review of Vision Science* 1, no. 1 (2015): 417–446.

Krizhevsky, Alex, Ilya Sutskever, and Geoffrey E. Hinton. "ImageNet Classification with Deep Convolutional Neural Networks." In *Proceedings of the 25th International Conference on Neural Information Processing Systems*. Vol. 1, 1097–1105. Lake Tahoe, Nevada: Curran Associates Inc., 2012. http://dl.acm.org/citation.cfm?id=2999134.2999257.

Krubitzer, Leah. "The Magnificent Compromise: Cortical Field Evolution in Mammals." *Neuron* 56, no. 2 (2007): 201–208.

Kruger, J., and D. Dunning. "Unskilled and Unaware of It: How Difficulties in Recognizing One's Own Incompetence Lead to Inflated

Self-Assessments." *Journal of Personality and Social Psychology* 77, no. 6 (1999): 1121–1134.

Krupenye, Christopher, and Josep Call. "Theory of Mind in Animals: Current and Future Directions." *WIREs Cognitive Science* 10, no. 6 (2019): e1503.

Kulke, Louisa, and Hannes Rakoczy. "Implicit Theory of Mind—an Overview of Current Replications and Non-Replications." *Data in Brief* 16 (2017): 101–104.

La Berge, S. P., L. E. Nagel, W. C. Dement, and V. P. Zarcone. "Lucid Dreaming Verified by Volitional Communication During REM Sleep." *Perceptual and Motor Skills* 52, no. 3 (1981): 727–732.

Lak, Armin, Gil M. Costa, Erin Romberg, Alexei A. Koulakov, Zachary F. Mainen, and Adam Kepecs. "Orbitofrontal Cortex Is Required for Optimal Waiting Based on Decision Confidence." *Neuron* 84, no. 1 (2014): 190–201.

Landman, Rogier, Henk Spekreijse, and Victor A. F. Lamme. "Large Capacity Storage of Integrated Objects Before Change Blindness." *Vision Research* 43, no. 2 (2003): 149–164.

Lau, Hakwan, and David Rosenthal. "Empirical Support for Higher-Order Theories of Conscious Awareness." *Trends in Cognitive Sciences* 15, no. 8 (2011): 365–373.

Lau, Hakwan, and R. E. Passingham. "Relative Blindsight in Normal Observers and the Neural Correlate of Visual Consciousness." *Proceedings of the National Academy of Sciences* 103, no. 49 (2006): 18763–18768.

Leary, Mark R. *The Curse of the Self: Self-Awareness, Egotism, and the Quality of Human Life*. Oxford: Oxford University Press, 2007.

Leary, Mark R., Kate J. Diebels, Erin K. Davisson, Katrina P. Jongman-Sereno, Jennifer C. Isherwood, Kaitlin T. Raimi, Samantha A. Deffler, and Rick H. Hoyle. "Cognitive and Interpersonal Features of Intellectual Humility." *Personality and Social Psychology Bulletin* 43, no. 6 (2017): 793–813.

LeDoux, Joseph. *Anxious: Using the Brain to Understand and Treat Fear and Anxiety*. 2015. Reprint, New York: Penguin, 2016.

LeDoux, Joseph E., and Richard Brown. "A Higher-Order Theory of Emotional Consciousness." *Proceedings of the National Academy of Science* 114, no. 10 (2017): E2016–E2025.

Lemaitre, Anne-Laure, Guillaume Herbet, Hugues Duffau, and Gilles Lafargue. "Preserved Metacognitive Ability Despite Unilateral or Bilateral Anterior Prefrontal Resection." *Brain and Cognition* 120 (2018): 48–57.

Lerch, Jason P., Adelaide P. Yiu, Alonso Martinez-Canabal, Tetyana Pekar, Veronique D. Bohbot, Paul W. Frankland, R. Mark Henkelman, Sheena A. Josselyn, and John G. Sled. "Maze Training in Mice Induces MRI-Detectable Brain Shape Changes Specific to the Type of Learning." *NeuroImage* 54, no. 3 (2011): 2086–2095.

Leslie, John. *The Philosophy of Arithmetic: Exhibiting a Progressive View of the Theory and Practice of Calculation, with an Enlarged Table of the Products of Numbers Under One Hundred*. Edinburgh: Constable and Co., 1817.

Lewis, Michael, and Douglas Ramsay. "Development of Self-Recognition, Personal Pronoun Use, and Pretend Play During the 2nd Year." *Child Development* 75, no. 6 (2004): 1821–1831.

Linnaeus, Carl. *Systema naturae, sive regna tria naturae systematice proposita per classes, ordines, genera, & species*. Lugdunum Batavorum, Netherlands, 1735.

Lockhart, Kristi L., Mariel K. Goddu, Eric D. Smith, and Frank C. Keil. "What Could You Really Learn on Your Own?: Understanding the Epistemic Limitations of Knowledge Acquisition." *Child Development* 87, no. 2 (2016): 477–493.

Lockl, Kathrin, and Wolfgang Schneider. "Knowledge About the Mind: Links Between Theory of Mind and Later Metamemory." *Child Development* 78, no. 1 (2007): 148–167.

Logan, Gordon D., and Matthew J. C. Crump. "Cognitive Illusions of Authorship Reveal Hierarchical Error Detection in Skilled Typists." *Science* 330, no. 6004 (2010): 683–686.

———. "Hierarchical Control of Cognitive Processes: The Case for Skilled Typewriting." In *The Psychology of Learning and Motivation: Advances in Research and Theory*. Vol. 54, edited by Brian H. Ross, 1–27. Cambridge, MA: Academic Press, 2011.

———. "The Left Hand Doesn't Know What the Right Hand Is Doing: The Disruptive Effects of Attention to the Hands in Skilled Typewriting." *Psychological Science* 20, no. 10 (2009): 1296–1300.

Logan, Gordon D., and N. Jane Zbrodoff. "Stroop-Type Interference: Congruity Effects in Color Naming with Typewritten Responses." *Journal of Experimental Psychology: Human Perception and Performance* 24, no. 3 (1998): 978–992.

Lou, Hans C., J. P. Changeux, and A. Rosenstand. "Towards a Cognitive Neuroscience of Self-Awareness." *Neuroscience & Biobehavioral Reviews* 83 (2017): 765–773.

Lou, Hans C., Bruce Luber, Michael Crupain, Julian P. Keenan, Markus Nowak, Troels W. Kjaer, Harold A. Sackeim, and Sarah H. Lisanby. "Parietal Cortex and Representation of the Mental Self." *Proceedings of the National Academy of Sciences* 101, no. 17 (2004): 6827–6832.

Ma, W. J., J. M. Beck, P. E. Latham, and A. Pouget. "Bayesian Inference with Probabilistic Population Codes." *Nature Neuroscience* 9, no. 11 (2006): 1432–1438.

Maguire, E. A., D. G. Gadian, I. S. Johnsrude, C. D. Good, J. Ashburner, R. S. Frackowiak, and Chris D. Frith. "Navigation-Related Structural Change in the Hippocampi of Taxi Drivers." *Proceedings of the National Academy of Sciences* 97, no. 8 (2000): 4398–4403.

Mandler, George. *A History of Modern Experimental Psychology: From James and Wundt to Cognitive Science*. Cambridge, MA: MIT Press, 2011.

Maniscalco, Brian, and Hakwan Lau. "A Signal Detection Theoretic Approach for Estimating Metacognitive Sensitivity from Confidence Ratings." *Consciousness and Cognition* 21, no. 1 (2012): 422–430.

Mansouri, Farshad Alizadeh, Etienne Koechlin, Marcello G. P. Rosa, and Mark J. Buckley. "Managing Competing Goals—a Key Role for the Frontopolar Cortex." *Nature Reviews Neuroscience* 18, no. 11 (2017): 645.

Marcus, Gary, and Ernest Davis. *Rebooting AI: Building Artificial Intelligence We Can Trust*. New York: Pantheon, 2019.

Margulies, Daniel S., Satrajit S. Ghosh, Alexandros Goulas, Marcel Falkiewicz, Julia M. Huntenburg, Georg Langs, Glen Bezgin et al. "Situating the Default-Mode Network Along a Principal Gradient of Macroscale Cortical Organization." *Proceedings of the National Academy of Sciences* 113, no. 44 (2016): 12574–12579.

Marr, D., and T. Poggio. "From Understanding Computation to Understanding Neural Circuitry." Massachusetts Institute of Technology Artificial Intelligence Laboratory, AI Memo No. 357, 1976.

Marsh, Henry. *Admissions: A Life in Brain Surgery*. London: Weidenfeld & Nicolson, 2017.

Mayer, Andreas, and Birgit E. Träuble. "Synchrony in the Onset of Mental State Understanding Across Cultures? A Study Among Children in Samoa." *International Journal of Behavioral Development* 37, no. 1 (2013): 21–28.

Mazancieux, Audrey, Stephen M. Fleming, Céline Souchay, and Chris J. A. Moulin. "Is There a G Factor for Metacognition? Correlations in Retrospective Metacognitive Sensitivity Across Tasks." *Journal of Experimental Psychology: General* 149, no. 9 (2020): 1788–1799.

McBrearty, Sally, and Alison S. Brooks. "The Revolution That Wasn't: A New Interpretation of the Origin of Modern Human Behavior." *Journal of Human Evolution* 39, no. 5 (2000): 453–563.

McCurdy, Li Yan, Brian Maniscalco, Janet Metcalfe, Ka Yuet Liu, Floris P. de Lange, and Hakwan Lau. "Anatomical Coupling Between Distinct Metacognitive Systems for Memory and Visual Perception." *Journal of Neuroscience* 33, no. 5 (2013): 1897–1906.

McGrayne, Sharon Bertsch. *The Theory That Would Not Die: How Bayes' Rule Cracked the Enigma Code, Hunted Down Russian Submarines, and Emerged Triumphant from Two Centuries of Controversy*. New Haven, CT: Yale University Press, 2012.

McGurk, Harry, and John MacDonald. "Hearing Lips and Seeing Voices." *Nature* 264, no. 5588 (1976): 746–748.

Meckler, Cédric, Laurence Carbonnell, Céline Ramdani, Thierry Hasbroucq, and Franck Vidal. "On-Line Action Monitoring of Response Execution:

An Electrophysiological Study." *Biological Psychology* 129 (2017): 178–185.

Mesulam, M. M. "From Sensation to Cognition." *Brain* 121, no. 6 (1998): 1013–1052.

Metcalfe, Janet. "Metacognitive Judgments and Control of Study." *Current Directions in Psychological Science*, June 1, 2009.

Metcalfe, Janet, and Brigid Finn. "Evidence That Judgments of Learning Are Causally Related to Study Choice." *Psychonomic Bulletin & Review* 15, no. 1 (2008): 174–179.

Metcalfe, Janet, and Nate Kornell. "The Dynamics of Learning and Allocation of Study Time to a Region of Proximal Learning." *Journal of Experimental Psychology: General* 132, no. 4 (2003): 530–542.

———. "A Region of Proximal Learning Model of Study Time Allocation." *Journal of Memory and Language* 52, no. 4 (2005): 463–477.

Metcalfe, Janet, and Arthur P. Shimamura, eds. *Metacognition: Knowing About Knowing*. 1994. Reprint, Cambridge, MA: MIT Press, 1996.

Metcalfe, Janet, and Lisa K. Son. "Anoetic, Noetic, and Autonoetic Metacognition." In *Foundations of Metacognition*, edited by Michael Beran, Johannes Brandl, Josef Perner, and Joëlle Proust. Oxford: Oxford University Press, 2012.

Metcalfe, Janet, Jared X. van Snellenberg, Pamela Derosse, Peter Balsam, and Anil K. Malhotra. "Judgements of Agency in Schizophrenia: An Impairment in Autonoetic Metacognition." *Philosophical Transactions of the Royal Society B: Biological Sciences* 367, no. 1594 (2012): 1391–1400.

Metzinger, Thomas. "M-Autonomy." *Journal of Consciousness Studies* 22, nos. 11–12 (2015): 270–302.

Meyniel, Florent, Daniel Schlunegger, and Stanislas Dehaene. "The Sense of Confidence During Probabilistic Learning: A Normative Account." *PLOS Computational Biology* 11, no. 6 (2015): e1004305.

Meyniel, Florent, Mariano Sigman, and Zachary F. Mainen. "Confidence as Bayesian Probability: From Neural Origins to Behavior." *Neuron* 88, no. 1 (2015): 78–92.

Michalsky, Tova, Zemira R. Mevarech, and Liora Haibi. "Elementary School Children Reading Scientific Texts: Effects of Metacognitive Instruction." *Journal of Educational Research* 102, no. 5 (2009): 363–376.

Michel, Matthias, and Jorge Morales. "Minority Reports: Consciousness and the Prefrontal Cortex." *Mind & Language* 35, no. 4 (2020): 493–513.

Middlebrooks, Paul G., and Marc A. Sommer. "Neuronal Correlates of Metacognition in Primate Frontal Cortex." *Neuron* 75, no. 3 (2012): 517–530.

Mill, John Stuart. *Auguste Comte and Positivism: Reprinted from the Westminster Review*. London: N. Trübner, 1865.

Mirels, Herbert L., Paul Greblo, and Janet B. Dean. "Judgmental Self-Doubt: Beliefs About One's Judgmental Prowess." *Personality and Individual Differences* 33, no. 5 (2002): 741–758.

Mitchell, Jason P., C. Neil Macrae, and Mahzarin R. Banaji. "Dissociable Medial Prefrontal Contributions to Judgments of Similar and Dissimilar Others." *Neuron* 50, no. 4 (2006): 655–663.

Miyamoto, Kentaro, Takahiro Osada, Rieko Setsuie, Masaki Takeda, Keita Tamura, Yusuke Adachi, and Yasushi Miyashita. "Causal Neural Network of Metamemory for Retrospection in Primates." *Science* 355, no. 6321 (2017): 188–193.

Miyamoto, Kentaro, Rieko Setsuie, Takahiro Osada, and Yasushi Miyashita. "Reversible Silencing of the Frontopolar Cortex Selectively Impairs Metacognitive Judgment on Non-experience in Primates." *Neuron* 97, no. 4 (2018): 980–989.e6.

Modirrousta, Mandana, and Lesley K. Fellows. "Medial Prefrontal Cortex Plays a Critical and Selective Role in 'Feeling of Knowing' Meta-Memory Judgments." *Neuropsychologia* 46, no. 12 (2008): 2958–2965.

Moeller, Scott J., and Rita Z. Goldstein. "Impaired Self-Awareness in Human Addiction: Deficient Attribution of Personal Relevance." *Trends in Cognitive Sciences* 18, no. 12 (2014).

Moore, James W., David Lagnado, Darvany C. Deal, and Patrick Haggard. "Feelings of Control: Contingency Determines Experience of Action." *Cognition* 110, no. 2 (2009): 279–283.

Morales, Jorge, Hakwan Lau, and Stephen M. Fleming. "Domain-General and Domain-Specific Patterns of Activity Supporting Metacognition in Human Prefrontal Cortex." *Journal of Neuroscience* 38, no. 14 (2018): 3534–3546.

Moritz, Steffen, and Todd S. Woodward. "Metacognitive Training in Schizophrenia: From Basic Research to Knowledge Translation and Intervention." *Current Opinion in Psychiatry* 20, no. 6 (2007): 619–625.

Moritz, Steffen, Christina Andreou, Brooke C. Schneider, Charlotte E. Wittekind, Mahesh Menon, Ryan P. Balzan, and Todd S. Woodward. "Sowing the Seeds of Doubt: A Narrative Review on Metacognitive Training in Schizophrenia." *Clinical Psychology Review* 34, no. 4 (2014): 358–366.

Morris, Robin G., and Daniel C. Mograbi. "Anosognosia, Autobiographical Memory and Self Knowledge in Alzheimer's Disease." *Cortex* 49, no. 6 (2013): 1553–1565.

Moulin, Chris J. A., Timothy J. Perfect, and Roy W. Jones. "Evidence for Intact Memory Monitoring in Alzheimer's Disease: Metamemory Sensitivity at Encoding." *Neuropsychologia* 38, no. 9 (2000): 1242–1250.

Mueller, Pam A., and Daniel M. Oppenheimer. "The Pen Is Mightier than the Keyboard: Advantages of Longhand over Laptop Note Taking." *Psychological Science* 25, no. 6 (2014): 1159–1168.

National Research Council. *Identifying the Culprit: Assessing Eyewitness Identification.* Washington, DC: National Academies Press, 2015.

Nelson, T. O. "A Comparison of Current Measures of the Accuracy of Feeling-of-Knowing Predictions." *Psychological Bulletin* 95 (1984): 109–133.

Nelson, T. O., J. Dunlosky, D. M. White, J. Steinberg, B. D. Townes, and D. Anderson. "Cognition and Metacognition at Extreme Altitudes on Mount Everest." *Journal of Experimental Psychology: General* 119, no. 4 (1990): 367–374.

Nelson, T. O., and L. Narens. "Metamemory: A Theoretical Framework and New Findings." In *Psychology of Learning and Motivation: Advances in Research and Theory.* Vol. 26, edited by Gordon H. Bower, 125–173. Cambridge, MA: Academic Press, 1990.

Nestor, James. *Deep: Freediving, Renegade Science, and What the Ocean Tells Us About Ourselves.* Boston: Eamon Dolan, 2014.

Neubert, Franz-Xaver, Rogier B. Mars, Adam G. Thomas, Jérôme Sallet, and Matthew F. S. Rushworth. "Comparison of Human Ventral Frontal Cortex Areas for Cognitive Control and Language with Areas in Monkey Frontal Cortex." *Neuron* 81, no. 3 (2014): 700–713.

Nicholson, Toby, David M. Williams, Catherine Grainger, Sophie E. Lind, and Peter Carruthers. "Relationships Between Implicit and Explicit Uncertainty Monitoring and Mindreading: Evidence from Autism Spectrum Disorder." *Consciousness and Cognition* 70 (2019): 11–24.

Nicholson, Toby, David M. Williams, Sophie E. Lind, Catherine Grainger, and Peter Carruthers. "Linking Metacognition and Mindreading: Evidence from Autism and Dual-Task Investigations." *Journal of Experimental Psychology: General*, September 10, 2020 (epub ahead of print).

Nieuwenhuis, S., K. R. Ridderinkhof, J. Blom, G. P. H. Band, and A. Kok. "Error-Related Brain Potentials Are Differentially Related to Awareness of Response Errors: Evidence from an Antisaccade Task." *Psychophysiology* 38, no. 5 (2001): 752–760.

Nisbett, R. E., and T. D. Wilson. "Telling More Than We Can Know: Verbal Reports on Mental Processes." *Psychological Review* 84, no. 3 (1977): 231.

Norman, Donald A., and Tim Shallice. "Attention to Action." In *Consciousness and Self-Regulation*, 1–18. Boston: Springer, 1986.

Northoff, G., A. Heinzel, M. de Greck, F. Bermpohl, H. Dobrowolny, and J. Panksepp. "Self-Referential Processing in Our Brain—a Meta-Analysis of Imaging Studies on the Self." *Neuroimage* 31, no. 1 (2006): 440–457.

O'Doherty, J. P., P. Dayan, K. Friston, H. Critchley, and R. J. Dolan. "Temporal Difference Models and Reward-Related Learning in the Human Brain." *Neuron* 38, no. 2 (2003): 329–337.

Ochsner, Kevin N., Kyle Knierim, David H. Ludlow, Josh Hanelin, Tara Ramachandran, Gary Glover, and Sean C. Mackey. "Reflecting upon Feelings: An fMRI Study of Neural Systems Supporting the Attribution of Emotion to Self and Other." *Journal of Cognitive Neuroscience* 16, no. 10 (2004): 1746–1772.

Olkowicz, Seweryn, Martin Kocourek, Radek K. Lučan, Michal Porteš, W. Tecumseh Fitch, Suzana Herculano-Houzel, and Pavel Němec. "Birds Have Primate-Like Numbers of Neurons in the Forebrain." *Proceedings of the National Academy of Sciences* 113, no. 26 (2016): 7255–7260.

Onishi, Kristine H., and Renée Baillargeon. "Do 15-Month-Old Infants Understand False Beliefs?" *Science* 308, no. 5719 (2005): 255–258.

Open Science Collaboration. "Estimating the Reproducibility of Psychological Science." *Science* 349, no. 6251 (2015).

Ortoleva, Pietro, and Erik Snowberg. "Overconfidence in Political Behavior." *American Economic Review* 105, no. 2 (2015): 504–535.

Palser, E. R., A. Fotopoulou, and J. M. Kilner. "Altering Movement Parameters Disrupts Metacognitive Accuracy." *Consciousness and Cognition* 57 (2018): 33–40.

Panagiotaropoulos, Theofanis I., Gustavo Deco, Vishal Kapoor, and Nikos K. Logothetis. "Neuronal Discharges and Gamma Oscillations Explicitly Reflect Visual Consciousness in the Lateral Prefrontal Cortex." *Neuron* 74, no. 5 (2012): 924–935.

Pannu, J. K., and A. W. Kaszniak. "Metamemory Experiments in Neurological Populations: A Review." *Neuropsychology Review* 15, no. 3 (2005): 105–130.

Park, JaeHong, Prabhudev Konana, Bin Gu, Alok Kumar, and Rajagopal Raghunathan. "Confirmation Bias, Overconfidence, and Investment Performance: Evidence from Stock Message Boards." *McCombs Research Paper Series* No. IROM-07-10 (2010).

Pasquali, Antoine, Bert Timmermans, and Axel Cleeremans. "Know Thyself: Metacognitive Networks and Measures of Consciousness." *Cognition* 117, no. 2 (2010): 182–190.

Passingham, R. E., S. L. Bengtsson, and Hakwan Lau. "Medial Frontal Cortex: From Self-Generated Action to Reflection on One's Own Performance." *Trends in Cognitive Sciences* 14, no. 1 (2010): 16–21.

Passingham, R. E., and Hakwan Lau. "Acting, Seeing, and Conscious Awareness." *Neuropsychologia* 128 (2019): 241–248.

Patel, D., Stephen M. Fleming, and J. M. Kilner. "Inferring Subjective States Through the Observation of Actions." *Proceedings of the Royal Society B: Biological Sciences* 279, no. 1748 (2012): 4853–4860.

Paulus, Markus, Joëlle Proust, and Beate Sodian. "Examining Implicit Metacognition in 3.5-Year-Old Children: An Eye-Tracking and Pupillometric Study." *Frontiers in Psychology* 4 (2013): 145.

Peirce, Charles Sanders, and Joseph Jastrow. "On Small Differences in Sensation." *Memoirs of the National Academy of Sciences* 3 (1885): 73–83.

Pennycook, Gordon, Jonathan A. Fugelsang, and Derek J. Koehler. "Everyday Associations of Analytic Thinking." *Current Directions in Psychological Science* 24, no. 6 (2015): 425–432.

Pennycook, Gordon, and David G. Rand. "Lazy, Not Biased: Susceptibility to Partisan Fake News Is Better Explained by Lack of Reasoning Than by Motivated Reasoning." *Cognition* 188 (2019): 39–50.

Pereira, Michael, Nathan Faivre, Iñaki Iturrate, Marco Wirthlin, Luana Serafini, Stéphanie Martin, Arnaud Desvachez, Olaf Blanke, Dimitri van de Ville, and José del R Millán. "Disentangling the Origins of Confidence in Speeded Perceptual Judgments Through Multimodal Imaging." *Proceedings of the National Academy of Sciences* 117, no. 15 (2020): 8382–8390.

Perner, Josef. "MiniMeta: In Search of Minimal Criteria for Metacognition." In *Foundations of Metacognition*, edited by Michael J. Beran, Johannes Brandl, Josef Perner, and Joëlle Proust, 94–116. Oxford: Oxford University Press, 2012.

Persaud, Navindra, Matthew Davidson, Brian Maniscalco, Dean Mobbs, R. E. Passingham, Alan Cowey, and Hakwan Lau. "Awareness-Related Activity in Prefrontal and Parietal Cortices in Blindsight Reflects More than Superior Visual Performance." *NeuroImage* 58, no. 2 (2011): 605–611.

Persaud, Navindra, P. McLeod, and A. Cowey. "Post-Decision Wagering Objectively Measures Awareness." *Nature Neuroscience* 10, no. 2 (2007): 257–261.

Peters, Megan A. K., Thomas Thesen, Yoshiaki D. Ko, Brian Maniscalco, Chad Carlson, Matt Davidson, Werner Doyle, Ruben Kuzniecky, Orrin Devinsky, Eric Halgren, and Hakwan Lau. "Perceptual Confidence Neglects Decision-Incongruent Evidence in the Brain." *Nature Human Behaviour 1*, no. 7 (2017): 0139.

Pezzulo, Giovanni, Francesco Rigoli, and Karl Friston. "Active Inference, Homeostatic Regulation and Adaptive Behavioural Control." *Progress in Neurobiology* 134 (2015): 17–35.

Phillips, Ian. "Blindsight Is Qualitatively Degraded Conscious Vision." *Psychological Review*, August 6, 2020 (epub ahead of print).

———. "The Methodological Puzzle of Phenomenal Consciousness." *Philosophical Transactions of the Royal Society B: Biological Sciences* 373, no. 1755 (2018): 20170347.

Pick, Herbert L., David H. Warren, and John C. Hay. "Sensory Conflict in Judgments of Spatial Direction." *Perception & Psychophysics* 6, no. 4 (1969): 203–205.

Pitt, David. "Mental Representation." In *The Stanford Encyclopedia of Philosophy*, edited by Edward N. Zalta. Stanford, CA: Metaphysics Research Lab, Stanford University, Winter 2018. https://plato.stanford.edu/archives/win2018/entries/mental-representation/.

Pittampalli, Al. *Persuadable: How Great Leaders Change Their Minds to Change the World*. New York: Harper Business, 2016.

Poldrack, Russell A., Chris I. Baker, Joke Durnez, Krzysztof J. Gorgolewski, Paul M. Matthews, Marcus R. Munafò, Thomas E. Nichols,

Jean-Baptiste Poline, Edward Vul, and Tal Yarkoni. "Scanning the Horizon: Towards Transparent and Reproducible Neuroimaging Research." *Nature Reviews Neuroscience* 18, no. 2 (2017): 115–126.

Premack, David, and Guy Woodruff. "Does the Chimpanzee Have a Theory of Mind?" *Behavioral and Brain Sciences* 1, no. 4 (1978): 515–526.

Programme for International Student Assessment. *Results: Ready to Learn: Students' Engagement, Drive and Self-Beliefs*. Paris: OECD Publishing, 2013.

Proust, Joëlle. *The Philosophy of Metacognition: Mental Agency and Self-Awareness*. Oxford: Oxford University Press, 2013.

Pulford, Briony D., Andrew M. Colman, Eike K. Buabang, and Eva M. Krockow. "The Persuasive Power of Knowledge: Testing the Confidence Heuristic." *Journal of Experimental Psychology: General* 147, no. 10 (2018): 1431.

Pyers, Jennie E., and Ann Senghas. "Language Promotes False-Belief Understanding." *Psychological Science* 20, no. 7 (2009): 805–812.

Qiu, Lirong, Jie Su, Yinmei Ni, Yang Bai, Xuesong Zhang, Xiaoli Li, and Xiaohong Wan. "The Neural System of Metacognition Accompanying Decision-Making in the Prefrontal Cortex." *PLOS Biology* 16, no. 4 (2018): e2004037.

Rabbitt, P. "Error Correction Time Without External Error Signals." *Nature* 212, no. 5060 (1966): 438.

Rabbitt, P., and B. Rodgers. "What Does a Man Do After He Makes an Error? An Analysis of Response Programming." *Quarterly Journal of Experimental Psychology* 29, no. 4 (1977): 727–743.

Ramnani, Narender, and Adrian M. Owen. "Anterior Prefrontal Cortex: Insights into Function from Anatomy and Neuroimaging." *Nature Reviews Neuroscience* 5, no. 3 (2004): 184–194.

Reber, Rolf, and Norbert Schwarz. "Effects of Perceptual Fluency on J udgments of Truth." *Consciousness and Cognition* 8, no. 3 (1999): 338–342.

Renz, Ursula, ed. *Self-Knowledge: A History*. Oxford: Oxford University Press, 2017.

Reyes, Gabriel, Jaime R. Silva, Karina Jaramillo, Lucio Rehbein, and Jérôme Sackur. "Self-Knowledge Dim-Out: Stress Impairs Metacognitive Accuracy." *PLOS One* 10, no. 8 (2015).

Reyes, Gabriel, Anastassia Vivanco-Carlevari, Franco Medina, Carolina Manosalva, Vincent de Gardelle, Jérôme Sackur, and Jaime R. Silva. "Hydrocortisone Decreases Metacognitive Efficiency Independent of Perceived Stress." *Scientific Reports* 10, no. 1 (2020): 1–9.

Reynolds, R. F., and A. M. Bronstein. "The Broken Escalator Phenomenon." *Experimental Brain Research* 151, no. 3 (2003): 301–308.

Ribas-Fernandes, J. J. F., A. Solway, C. Diuk, J. T. McGuire, A. G. Barto, Y. Niv, and M. M. Botvinick. "A Neural Signature of Hierarchical Reinforcement Learning." *Neuron* 71, no. 2 (2011): 370–379.

Risko, Evan F., and Sam J. Gilbert. "Cognitive Offloading." *Trends in Cognitive Sciences* 20, no. 9 (2016): 676–688.

Ro, Tony, Dominique Shelton, Olivia L. Lee, and Erik Chang. "Extragenic-ulate Mediation of Unconscious Vision in Transcranial Magnetic Stimulation-Induced Blindsight." *Proceedings of the National Academy of Sciences* 101, no. 26 (2004): 9933–9935.

Roca, Maria, Teresa Torralva, Ezequiel Gleichgerrcht, Alexandra Woolgar, Russell Thompson, John Duncan, and Facundo Manes. "The Role of Area 10 (Ba10) in Human Multitasking and in Social Cognition: A Lesion Study." *Neuropsychologia* 49, no. 13 (2011): 3525–3531.

Rohrer, Julia, Warren Tierney, Eric L. Uhlmann, Lisa M. DeBruine, Tom Heymann, Benedict Jones, Stefan C. Schmukle, et al. "Putting the Self in Self-Correction: Findings from the Loss-of-Confidence Project." *Perspectives on Psychological Science*, in press.

Rohwer, Michael, Daniela Kloo, and Josef Perner. "Escape from Metaigno-rance: How Children Develop an Understanding of Their Own Lack of Knowledge." *Child Development* 83, no. 6 (2012): 1869–1883.

Rollwage, Max, Raymond J. Dolan, and Stephen M. Fleming. "Metacogni-tive Failure as a Feature of Those Holding Radical Beliefs." *Current Biology* 28, no. 24 (2018): 4014–4021.e8.

Rollwage, Max, and Stephen M. Fleming. "Confirmation Bias Is Adaptive When Coupled with Efficient Metacognition." *Philosophical Transactions of the Royal Society B: Biological Sciences*, in press.

Rollwage, Max, Alisa Loosen, Tobias U. Hauser, Rani Moran, Raymond J. Dolan, and Stephen M. Fleming. "Confidence Drives a Neural Confirmation Bias." *Nature Communications* 11, no. 1 (2020): 1–11.

Ronfard, Samuel, and Kathleen H. Corriveau. "Teaching and Preschoolers' Ability to Infer Knowledge from Mistakes." *Journal of Experimental Child Psychology* 150 (2016): 87–98.

Rosenblatt, F. "The Perceptron: A Probabilistic Model for Information Stor-age and Organization in the Brain." *Psychological Review* 65, no. 6 (1958): 386–408.

Rosenthal, David M. *Consciousness and Mind*. Oxford: Oxford University Press, 2005.

Rouault, Marion, Peter Dayan, and Stephen M. Fleming. "Forming Global Estimates of Self-Performance from Local Confidence." *Nature Communications* 10, no. 1 (2019): 1–11.

Rouault, Marion, and Stephen M. Fleming. "Formation of Global Self-Beliefs in the Human Brain." *Proceedings of the National Academy of Sciences* 117, no. 44 (2020): 27268–27276.

Rouault, Marion, Andrew McWilliams, Micah G. Allen, and Stephen M. Flem-ing. "Human Metacognition Across Domains: Insights from Individual Differences and Neuroimaging." *Personality Neuroscience* 1 (2018).

Rouault, Marion, Tricia Seow, Claire M. Gillan, and Stephen M. Fleming. "Psychiatric Symptom Dimensions Are Associated with Dissociable

Shifts in Metacognition but Not Task Performance." *Biological Psychiatry* 84, no. 6 (2018): 443–451.

Rounis, Elisabeth, Brian Maniscalco, John C. Rothwell, Richard E. Passingham, and Hakwan Lau. "Theta-Burst Transcranial Magnetic Stimulation to the Prefrontal Cortex Impairs Metacognitive Visual Awareness." *Cognitive Neuroscience* 1, no. 3 (2010): 165–175.

Rumelhart, David E., Geoffrey E. Hinton, and Ronald J. Williams. "Learning Representations by Back-Propagating Errors." *Nature* 323, no. 6088 (1986): 533–536.

Ryle, Gilbert. *The Concept of Mind.* Chicago: University of Chicago Press, 2012.

Sahraie, A., L. Weiskrantz, J. L. Barbur, A. Simmons, S. C. R. Williams, and M. J. Brammer. "Pattern of Neuronal Activity Associated with Conscious and Unconscious Processing of Visual Signals." *Proceedings of the National Academy of Sciences* 94, no. 17 (1997).

Samaha, Jason, Missy Switzky, and Bradley R. Postle. "Confidence Boosts Serial Dependence in Orientation Estimation." *Journal of Vision* 19, no. 4 (2019): 25.

Samek, Wojciech, Grégoire Montavon, Andrea Vedaldi, Lars Kai Hansen, and Klaus-Robert Müller. *Explainable AI: Interpreting, Explaining and Visualizing Deep Learning.* Cham, Switzerland: Springer, 2019.

Schäfer, Anton Maximilian, and Hans-Georg Zimmermann. "Recurrent Neural Networks Are Universal Approximators." *International Journal of Neural Systems* 17, no. 4 (2007): 253–263.

Schechtman, Marya. *The Constitution of Selves.* Ithaca, NY: Cornell University Press, 1996.

Schellings, Gonny L. M., Bernadette H. A. M. van Hout-Wolters, Marcel V. J. Veenman, and Joost Meijer. "Assessing Metacognitive Activities: The In-Depth Comparison of a Task-Specific Questionnaire with Think-Aloud Protocols." *European Journal of Psychology of Education* 28, no. 3 (2013): 963–990.

Schlerf, John E., Joseph M. Galea, Amy J. Bastian, and Pablo A. Celnik. "Dynamic Modulation of Cerebellar Excitability for Abrupt, but Not Gradual, Visuomotor Adaptation." *Journal of Neuroscience* 32, no. 34 (2012): 11610–11617.

Schmid, Michael C., Sylwia W. Mrowka, Janita Turchi, Richard C. Saunders, Melanie Wilke, Andrew J. Peters, Frank Q. Ye, and David A. Leopold. "Blindsight Depends on the Lateral Geniculate Nucleus." *Nature* 466, no. 7304 (2010): 373–377.

Schmidt, Carlos, Gabriel Reyes, Mauricio Barrientos, Álvaro I. Langer, and Jérôme Sackur. "Meditation Focused on Self-Observation of the Body Impairs Metacognitive Efficiency." *Consciousness and Cognition* 70 (2019): 116–125.

Schmitz, Taylor W., Howard A. Rowley, Tisha N. Kawahara, and Sterling C. Johnson. "Neural Correlates of Self-Evaluative Accuracy After Traumatic Brain Injury." *Neuropsychologia* 44, no. 5 (2006): 762–773.

Schneider, Peter, Michael Scherg, H. Günter Dosch, Hans J. Specht, Alexander Gutschalk, and André Rupp. "Morphology of Heschl's Gyrus Reflects Enhanced Activation in the Auditory Cortex of Musicians." *Nature Neuroscience* 5, no. 7 (2002): 688–694.

Schnyer, David M., Mieke Verfaellie, Michael P. Alexander, Ginette LaFleche, Lindsay Nicholls, and Alfred W Kaszniak. "A Role for Right Medial Prefrontal Cortex in Accurate Feeling-of-Knowing Judgements: Evidence from Patients with Lesions to Frontal Cortex." *Neuropsychologia* 42, no. 7 (2004): 957–966.

Scholz, Jan, Miriam C. Klein, Timothy E. J. Behrens, and Heidi Johansen-Berg. "Training Induces Changes in White-Matter Architecture." *Nature Neuroscience* 12, no. 11 (2009): 1370–1371.

Schooler, Jonathan W. "Re-Representing Consciousness: Dissociations Between Experience and Meta-Consciousness." *Trends in Cognitive Sciences* 6, no. 8 (2002): 339–344.

Schooler, Jonathan W., Jonathan Smallwood, Kalina Christoff, Todd C. Handy, Erik D. Reichle, and Michael A. Sayette. "Meta-Awareness, Perceptual Decoupling and the Wandering Mind." *Trends in Cognitive Sciences* 15, no. 7 (2011): 319–326.

Schultz, W., P. Dayan, and P. R. Montague. "A Neural Substrate of Prediction and Reward." *Science* 275, no. 5306 (1997): 1593.

Schulz, Lion, Max Rollwage, Raymond J. Dolan, and Stephen M. Fleming. "Dogmatism Manifests in Lowered Information Search Under Uncertainty." *Proceedings of the National Academy of Sciences*, 117, no. 49 (2020): 31527–31534.

Schurger, Aaron, Steven Gale, Olivia Gozel, and Olaf Blanke. "Performance Monitoring for Brain-Computer-Interface Actions." *Brain and Cognition* 111 (2017): 44–50.

Scott, Rose M., and Renée Baillargeon. "Early False-Belief Understanding." *Trends in Cognitive Sciences* 21, no. 4 (2017): 237–249.

Semendeferi, Katerina, Kate Teffer, Dan P. Buxhoeveden, Min S. Park, Sebastian Bludau, Katrin Amunts, Katie Travis, and Joseph Buckwalter. "Spatial Organization of Neurons in the Frontal Pole Sets Humans Apart from Great Apes." *Cerebral Cortex* 21, no. 7 (2010): 1485–1497.

Seth, Anil K. "Interoceptive Inference, Emotion, and the Embodied Self." *Trends in Cognitive Sciences* 17, no. 11 (2013): 565–573.

Seymour, Ben, John P. O'Doherty, Peter Dayan, Martin Koltzenburg, Anthony K. Jones, Raymond J. Dolan, Karl J. Friston, and Richard S. Frackowiak. "Temporal Difference Models Describe Higher-Order Learning in Humans." *Nature* 429, no. 6992 (2004): 664–667.

Shea, Nicholas. *Representation in Cognitive Science*. Oxford: Oxford University Press, 2018.

Shea, Nicholas, Annika Boldt, Dan Bang, Nick Yeung, Cecilia Heyes, and Chris D. Frith. "Supra-Personal Cognitive Control and Metacognition." *Trends in Cognitive Sciences* 18, no. 4 (2014): 186–193.

Shekhar, Medha, and Dobromir Rahnev. "Distinguishing the Roles of Dorsolateral and Anterior PFC in Visual Metacognition." *Journal of Neuroscience* 38, no. 22 (2018): 5078–5087.

———. "The Nature of Metacognitive Inefficiency in Perceptual Decision-Making." *Psychological Review* 128, no. 1 (2021): 45.

Shergill, Sukhwinder S., Paul M. Bays, Chris D. Frith, and Daniel M. Wolpert. "Two Eyes for an Eye: The Neuroscience of Force Escalation." *Science* 301, no. 5630 (2003): 187.

Shidara, Munetaka, and Barry J. Richmond. "Anterior Cingulate: Single Neuronal Signals Related to Degree of Reward Expectancy." *Science* 296, no. 5573 (2002): 1709–1711.

Shields, Wendy E., J. David Smith, and David A. Washburn. "Uncertain Responses by Humans and Rhesus Monkeys (Macaca mulatta) in a Psychophysical Same-Different Task." *Journal of Experimental Psychology: General* 126, no. 2 (1997): 147.

Shimamura, Arthur P. "Toward a Cognitive Neuroscience of Metacognition." *Consciousness and Cognition* 9, no. 2 (2000): 313–323.

Shimamura, Arthur P., and L. R. Squire. "Memory and Metamemory: A Study of the Feeling-of-Knowing Phenomenon in Amnesic Patients." *Journal of Experimental Psychology: Learning, Memory, and Cognition* 12, no. 3 (1986): 452–460.

Siedlecka, Marta, Boryslaw Paulewicz, and Michal Wierzchoń. "But I Was So Sure! Metacognitive Judgments Are Less Accurate Given Prospectively Than Retrospectively." *Frontiers in Psychology* 7, no. 240 (2016): 218.

Silver, David, Julian Schrittwieser, Karen Simonyan, Ioannis Antonoglou, Aja Huang, Arthur Guez, Thomas Hubert, et al. "Mastering the Game of Go Without Human Knowledge." *Nature* 550, no. 7676 (2017): 354–359.

Simons, Daniel J. "Unskilled and Optimistic: Overconfident Predictions Despite Calibrated Knowledge of Relative Skill." *Psychonomic Bulletin & Review* 20, no. 3 (2013): 601–607.

Simons, Jon S., Jane R. Garrison, and Marcia K. Johnson. "Brain Mechanisms of Reality Monitoring." *Trends in Cognitive Sciences* 21, no. 6 (2017): 462–473.

Simons, Jon S., Polly V. Peers, Yonatan S. Mazuz, Marian E. Berryhill, and Ingrid R. Olson. "Dissociation Between Memory Accuracy and Memory Confidence Following Bilateral Parietal Lesions." *Cerebral Cortex* 20, no. 2 (2010): 479–485.

Sinclair, Alyssa H., Matthew L. Stanley, and Paul Seli. "Closed-Minded Cognition: Right-Wing Authoritarianism Is Negatively Related to Belief Updating Following Prediction Error." *Psychonomic Bulletin & Review*, July 27, 2020, 1–14.

Smallwood, Jonathan, and Jonathan W. Schooler. "The Restless Mind." *Psychological Bulletin* 132, no. 6 (2006): 946–958.

Smith, J. David, Jonathan Schull, Jared Strote, Kelli McGee, Roian Egnor, and Linda Erb. "The Uncertain Response in the Bottlenosed Dolphin (Tursiops Truncatus)." *Journal of Experimental Psychology: General* 124, no. 4 (1995): 391–408.

Song, C., R. Kanai, Stephen M. Fleming, Rimona S. Weil, D. S. Schwarzkopf, and Geraint Rees. "Relating Inter-Individual Differences in Metacognitive Performance on Different Perceptual Tasks." *Consciousness and Cognition* 20, no. 4 (2011): 1787–1792.

Sperber, Dan, and Hugo Mercier. *The Enigma of Reason: A New Theory of Human Understanding.* London: Allen Lane, 2017.

Sperling, George. "The Information Available in Brief Visual Presentations." *Psychological Monographs: General and Applied* 74, no. 11 (1960): 1–29.

Stazicker, James. "Partial Report Is the Wrong Paradigm." *Philosophical Transactions of the Royal Society B: Biological Sciences* 373, no. 1755 (2018): 20170350.

Stephan, Klaas E., Zina M. Manjaly, Christoph D. Mathys, Lilian A. E. Weber, Saee Paliwal, Tim Gard, Marc Tittgemeyer, et al. "Allostatic Self-Efficacy: A Metacognitive Theory of Dyshomeostasis-Induced Fatigue and Depression." *Frontiers in Human Neuroscience* 10 (2016): 550.

Sterelny, Kim. "From Hominins to Humans: How *Sapiens* Became Behaviourally Modern." *Philosophical Transactions of the Royal Society B: Biological Sciences* 366, no. 1566 (2011): 809–822.

Sterling, Peter. "Allostasis: A Model of Predictive Regulation." *Physiology & Behavior* 106, no. 1 (2012): 5–15.

Stuss, D. T., M. P. Alexander, A. Lieberman, and H. Levine. "An Extraordinary Form of Confabulation." *Neurology* 28, no. 11 (1978): 1166–1172.

Summerfield, Jennifer J., Demis Hassabis, and Eleanor A. Maguire. "Cortical Midline Involvement in Autobiographical Memory." *NeuroImage* 44, no. 3 (2009): 1188–1200.

Sunstein, Cass R., Sebastian Bobadilla-Suarez, Stephanie C. Lazzaro, and Tali Sharot. "How People Update Beliefs About Climate Change: Good News and Bad News." *Cornell Law Review* 102 (2016): 1431.

Sutton, Richard S., and Andrew G. Barto. *Reinforcement Learning: An Introduction.* Cambridge, MA: MIT Press, 2018.

Talluri, Bharath Chandra, Anne E. Urai, Konstantinos Tsetsos, Marius Usher, and Tobias H. Donner. "Confirmation Bias Through Selective Overweighting of Choice-Consistent Evidence." *Current Biology* 28, no. 19 (2018): 3128–3135.

Tauber, Sarah K., and Matthew G. Rhodes. "Metacognitive Errors Contribute to the Difficulty in Remembering Proper Names." *Memory* 18, no. 5 (2010): 522–532.

Tegmark, Max. *Life 3.0: Being Human in the Age of Artificial Intelligence.* New York: Knopf, 2017.

Tetlock, Philip E., and Dan Gardner. *Superforecasting: The Art and Science of Prediction.* New York: Random House, 2016.

Thompson, Valerie A., Jamie A. Prowse Turner, Gordon Pennycook, Linden J. Ball, Hannah Brack, Yael Ophir, and Rakefet Ackerman. "The Role of Answer Fluency and Perceptual Fluency as Metacognitive Cues for Initiating Analytic Thinking." *Cognition* 128, no. 2 (2013): 237–251.

Thornton, Mark A., Miriam E. Weaverdyck, Judith N. Mildner, and Diana I. Tamir. "People Represent Their Own Mental States More Distinctly Than Those of Others." *Nature Communications* 10, no. 1 (2019): 2117.

Toner, Kaitlin, Mark R. Leary, Michael W. Asher, and Katrina P. Jongman-Sereno. "Feeling Superior Is a Bipartisan Issue: Extremity (Not Direction) of Political Views Predicts Perceived Belief Superiority." *Psychological Science*, October 4, 2013.

Toplak, Maggie E., Richard F. West, and Keith E. Stanovich. "The Cognitive Reflection Test as a Predictor of Performance on Heuristics-and-Biases Tasks." *Memory & Cognition* 39, no. 7 (2011): 1275.

Torrecillos, Flavie, Philippe Albouy, Thomas Brochier, and Nicole Malfait. "Does the Processing of Sensory and Reward-Prediction Errors Involve Common Neural Resources? Evidence from a Frontocentral Negative Potential Modulated by Movement Execution Errors." *Journal of Neuroscience* 34, no. 14 (2014): 4845–4856.

Trouche, Emmanuel, Petter Johansson, Lars Hall, and Hugo Mercier. "The Selective Laziness of Reasoning." *Cognitive Science* 40, no. 8 (2016): 2122–2136.

Tulving, Endel. "Memory and Consciousness." *Canadian Psychology/Psychologie canadienne* 26, no. 1 (1985): 1–12.

Turing, Alan Mathison. "On Computable Numbers, with an Application to the Entscheidungsproblem." *Proceedings of the London Mathematical Society* 42, no. 1 (1937): 230–265.

Ullsperger, Markus, Helga A. Harsay, Jan R. Wessel, and K. Richard Ridderinkhof. "Conscious Perception of Errors and Its Relation to the Anterior Insula." *Brain Structure & Function* 214, nos. 5–6 (2010): 629–643.

Vaccaro, Anthony G., and Stephen M. Fleming. "Thinking About Thinking: A Coordinate-Based Meta-Analysis of Neuroimaging Studies of Metacognitive Judgements." *Brain and Neuroscience Advances* 2 (2018): 2398212818810591.

Van Dam, Nicholas T., Marieke K. van Vugt, David R. Vago, Laura Schmalzl, Clifford D. Saron, Andrew Olendzki, Ted Meissner, et al. "Mind the Hype: A Critical Evaluation and Prescriptive Agenda for Research on Mindfulness and Meditation." *Perspectives on Psychological Science* 13, no. 1 (2018): 36–61.

Van den Berg, Ronald, Aspen H. Yoo, and Wei Ji Ma. "Fechner's Law in Metacognition: A Quantitative Model of Visual Working Memory Confidence." *Psychological Review* 124, no. 2 (2017): 197–214.

Van der Plas, Elisa, Anthony S. David, and Stephen M. Fleming. "Advice-Taking as a Bridge Between Decision Neuroscience and Mental Capacity." *International Journal of Law and Psychiatry* 67 (2019): 101504.

Vannini, Patrizia, Federico d'Oleire Uquillas, Heidi I. L. Jacobs, Jorge Sepulcre, Jennifer Gatchel, Rebecca E. Amariglio, Bernard Hanseeuw, et al. "Decreased Meta-Memory Is Associated with Early Tauopathy in Cognitively Unimpaired Older Adults." *NeuroImage: Clinical* 24 (2019): 102097.

Vilkki, Juhani, Antti Servo, and Outi Surma-aho. "Word List Learning and Prediction of Recall After Frontal Lobe Lesions." *Neuropsychology* 12, no. 2 (1998): 268.

Vilkki, Juhani, Outi Surma-aho, and Antti Servo. "Inaccurate Prediction of Retrieval in a Face Matrix Learning Task After Right Frontal Lobe Lesions." *Neuropsychology* 13, no. 2 (1999): 298.

Von Hippel, William, and Robert Trivers. "The Evolution and Psychology of Self-Deception." *Behavioral and Brain Sciences* 34, no. 1 (2011): 1–16.

Voss, Ursula, Romain Holzmann, Allan Hobson, Walter Paulus, Judith Koppehele-Gossel, Ansgar Klimke, and Michael A. Nitsche. "Induction of Self Awareness in Dreams Through Frontal Low Current Stimulation of Gamma Activity." *Nature Neuroscience* 17, no. 6 (2014): 810–812.

Vygotsky, Lev Semenovich. *Thought and Language*. Cambridge, MA: MIT Press, 1986.

Wald, A. "Sequential Tests of Statistical Hypotheses." *Annals of Mathematical Statistics* 16, no. 2 (1945): 117–186.

Walker, Mary Jean. "Neuroscience, Self-Understanding, and Narrative Truth." *AJOB Neuroscience* 3, no. 4 (2012): 63–74.

Wallis, Jonathan D. "Cross-Species Studies of Orbitofrontal Cortex and Value-Based Decision-Making." *Nature Neuroscience* 15, no. 1 (2011): 13–19.

Wang, Jane X., Zeb Kurth-Nelson, Dharshan Kumaran, Dhruva Tirumala, Hubert Soyer, Joel Z. Leibo, Demis Hassabis, and Matthew Botvinick. "Prefrontal Cortex as a Meta-Reinforcement Learning System." *Nature Neuroscience* 21, no. 6 (2018): 860.

Wegner, Daniel M. *The Illusion of Conscious Will*. Cambridge, MA: MIT Press, 2003.

Weil, Leonora G., Stephen M. Fleming, Iroise Dumontheil, Emma J. Kilford, Rimona S. Weil, Geraint Rees, Raymond J. Dolan, and Sarah-Jayne Blakemore. "The Development of Metacognitive Ability in Adolescence." *Consciousness and Cognition* 22, no. 1 (2013): 264–271.

Weinberg, Robert, Daniel Gould, and Allen Jackson. "Expectations and Performance: An Empirical Test of Bandura's Self-Efficacy Theory." *Journal of Sport and Exercise Psychology* 1, no. 4 (1979): 320–331.

Weiskrantz, L., E. K. Warrington, M. D. Sanders, and J. Marshall. "Visual Capacity in the Hemianopic Field Following a Restricted Occipital Ablation." *Brain: A Journal of Neurology* 97, no. 4 (1974): 709–728.

Wenke, Dorit, Stephen M. Fleming, and Patrick Haggard. "Subliminal Priming of Actions Influences Sense of Control over Effects of Action." *Cognition* 115, no. 1 (2010): 26–38.

Wiesmann, Charlotte Grosse, Angela D. Friederici, Tania Singer, and Nikolaus Steinbeis. "Two Systems for Thinking About Others' Thoughts in the Developing Brain." *Proceedings of the National Academy of Sciences* 117, no. 12 (2020): 6928–6935.

Will, Geert-Jan, Robb B. Rutledge, Michael Moutoussis, and Raymond J. Dolan. "Neural and Computational Processes Underlying Dynamic Changes in Self-Esteem." *eLife* 6 (2017): e28098.

Wimmer, H., and J. Perner. "Beliefs About Beliefs: Representation and Constraining Function of Wrong Beliefs in Young Children's Understanding of Deception." *Cognition* 13, no. 1 (1983): 103–128.

Winkielman, Piotr, and Jonathan W. Schooler. "Unconscious, Conscious, and Metaconscious in Social Cognition." In *Social Cognition: The Basis of Human Interaction*, 49–69. New York: Psychology Press, 2009.

Wixted, John T., and Gary L. Wells. "The Relationship Between Eyewitness Confidence and Identification Accuracy: A New Synthesis." *Psychological Science in the Public Interest* 18, no. 1 (2017): 10–65.

Wolpert, D. M., and R. C. Miall. "Forward Models for Physiological Motor Control." *Neural Networks* 9, no. 8 (1996): 1265–1279.

Woolgar, Alexandra, Alice Parr, Rhodri Cusack, Russell Thompson, Ian Nimmo-Smith, Teresa Torralva, Maria Roca, Nagui Antoun, Facundo Manes, and John Duncan. "Fluid Intelligence Loss Linked to Restricted Regions of Damage Within Frontal and Parietal Cortex." *Proceedings of the National Academy of Sciences*, 107, no. 33 (2010): 14899–14902.

Woollett, Katherine, and Eleanor A. Maguire. "Acquiring 'The Knowledge' of London's Layout Drives Structural Brain Changes." *Current Biology* 21, no. 24 (2011): 2109–2114.

Ye, Qun, Futing Zou, Hakwan Lau, Yi Hu, and Sze Chai Kwok. "Causal Evidence for Mnemonic Metacognition in Human Precuneus." *Journal of Neuroscience* 38, no. 28 (2018): 6379–6387.

Yeung, N., J. D. Cohen, and M. M. Botvinick. "The Neural Basis of Error Detection: Conflict Monitoring and the Error-Related Negativity." *Psychological Review* 111, no. 4 (2004): 931–959.

Yokoyama, Osamu, Naoki Miura, Jobu Watanabe, Atsushi Takemoto, Shinya Uchida, Motoaki Sugiura, Kaoru Horie, Shigeru Sato, Ryuta Kawashima, and Katsuki Nakamura. "Right Frontopolar Cortex Activity Correlates with Reliability of Retrospective Rating of Confidence in Short-Term Recognition Memory Performance." *Neuroscience Research* 68, no. 3 (2010): 199–206.

Yon, Daniel, Cecilia Heyes, and Clare Press. "Beliefs and Desires in the Predictive Brain." *Nature Communications* 11, no. 1 (2020): 1–4.

Young, Andrew G., and Andrew Shtulman. "Children's Cognitive Reflection Predicts Conceptual Understanding in Science and Mathematics." *Psychological Science* 31, no. 11 (2020): 1396–1408.

Zacharopoulos, George, Nicola Binetti, V. Walsh, and R. Kanai. "The Effect of Self-Efficacy on Visual Discrimination Sensitivity." *PLOS One* 9, no. 10 (2014): e109392.

Zatorre, Robert J., R. Douglas Fields, and Heidi Johansen-Berg. "Plasticity in Gray and White: Neuroimaging Changes in Brain Structure During Learning." *Nature Neuroscience* 15, no. 4 (2012): 528–536.

Zeki, S., and A. Bartels. "The Autonomy of the Visual Systems and the Modularity of Conscious Vision." *Philosophical Transactions of the Royal Society B: Biological Sciences* 353, no. 1377 (1998): 1911–1914.

Zimmerman, Barry J. "Self-Regulated Learning and Academic Achievement: An Overview." *Educational Psychologist* 25, no. 1 (1990): 3–17.

Zylberberg, Ariel, Pablo Barttfeld, and Mariano Sigman. "The Construction of Confidence in a Perceptual Decision." *Frontiers in Integrative Neuroscience* 6 (2012): 79.

INDEX

STEPHEN M. FLEMING is a Wellcome Trust/Royal Society Sir Henry Dale Fellow at the Department of Experimental Psychology and Principal Investigator at the Wellcome Centre for Human Neuroimaging, University College London, where he leads the Metacognition Group. He lives in London.